[意大利]圭多·卡尔达雷利 米凯莱·卡坦扎罗 著 李果 译

牛津通识读本·

网络

Networks
A Very Short Introduction

译林出版社

图书在版编目（CIP）数据

网络 /（意）圭多·卡尔达雷利，（意）米凯莱·卡坦扎罗著；李果译 . —南京：译林出版社，2018.12（2022.10 重印）
（牛津通识读本）
书名原文：Networks: A Very Short Introduction
ISBN 978-7-5447-7438-3

Ⅰ. ①网⋯　Ⅱ. ①圭⋯　②米⋯　③李⋯　Ⅲ. ①系统科学 - 研究　Ⅳ. ①N94

中国版本图书馆 CIP 数据核字（2018）第 147036 号

Copyright © Guido Caldarelli and Michele Catanzaro, 2012
Networks was originally published in English in 2012.
This Bilingual Edition is published by arrangement with Oxford University Press and is for sale in the People's Republic of China only, excluding Hong Kong SAR, Macau SAR and Taiwan, and may not be bought for export therefrom.
Chinese and English edition copyright © 2018 by Yilin Press, Ltd

著作权合同登记号　图字：10-2012-485 号

网络　[意大利] 圭多·卡尔达雷利　米凯莱·卡坦扎罗 ／著　李　果 ／译

责任编辑	许　丹
装帧设计	景秋萍
校　　对	季林巧
责任印制	董　虎

原文出版	Oxford University Press, 2012
出版发行	译林出版社
地　　址	南京市湖南路 1 号 A 楼
邮　　箱	yilin@yilin.com
网　　址	www.yilin.com
市场热线	025-86633278
排　　版	南京展望文化发展有限公司
印　　刷	江苏凤凰通达印刷有限公司
开　　本	890 毫米 ×1260 毫米　1/32
印　　张	8.125
插　　页	4
版　　次	2018 年 12 月第 1 版
印　　次	2022 年 10 月第 5 次印刷
书　　号	ISBN 978-7-5447-7438-3
定　　价	39.00 元

版权所有　·　侵权必究

译林版图书若有印装错误可向出版社调换。质量热线：025-83658316

序 言

狄增如

著名诗人北岛有一首名为《生活》的一字诗：网。

毫无疑问，我们的世界运行在有形或无形的网络之中，尤其是在信息技术飞速发展的今天。进入21世纪以来，技术进步和社会经济的全球化，把世界更加紧密地联结在一起，系统性和复杂性已经成为各领域核心问题的共性特征。

正像《网络》一书中指出的，系统性和复杂性的现象在我们的生活中比比皆是，大到全球性的金融危机、生态系统的稳定性，小到空手道俱乐部的分裂和足球队的场上表现，其突出表现是系统各单元或要素，通过相互关联和影响所展现出的涌现性行为：在没有中心控制和全局信息的情况下，仅仅通过个体之间的局域相互作用，系统就可以在一定条件下展现出宏观的时空或功能结构，在新的层次上涌现出具有整体性的性质和功能。在这里应该强调的是，非线性的关联和相互作用关系对于系统的涌现性至关重要，它使得系统的整体行为不能通过个体行为的简单叠加而获得。因此，结构决定系统功能，刻画系统中的相互作用关系对于认识和理解系统的宏观性质意义重大。

而复杂网络就是对系统结构的最一般和本质的刻画。系统中的个体对应于网络的节点,个体之间的联系或相互作用对应于联结节点的边。网络的抽象,超越了每个系统自身的特殊性质、独特背景和演化机制,让我们可以用统一、普适的网络分析方法去研究系统的性质,进而加深对系统共性的了解。事实上,自18世纪欧拉建立起图与网络的概念,解决了哥尼斯堡七桥问题开始,图与网络就成为刻画结构的重要工具,并于19世纪中叶进入社会学领域,社会网络分析由此成为社会计量学的重要工具。

1998年,沃茨-斯托加茨提出小世界网络模型,指出少量的随机捷径会改变网络的拓扑结构,从而涌现出小世界的效应;随后,巴拉巴西提出了无标度网络概念,解释了增长和择优机制在复杂网络自组织演化过程中的普遍性和幂律的重要性。这些开创性的研究工作使我们认识到,实际系统的网络结构存在着许多超越经典随机网络的特殊性质,例如小世界性质、幂律度分布、不同的匹配关系、社团结构等,我们需要建立更加细致、准确的概念,刻画网络结构并进而更加科学地认识系统性质,以促进网络科学的快速发展。

网络描述方法已被广泛应用于实际系统的研究,例如神经元网络、食物链网络、因特网络、赛博空间以及人与人之间交往的社会网络等等,研究结果加深了对这些具体系统的理解,并且提出了一系列新的概念和分析方法。总体上说,网络科学研究包括以下主要内容:一、如何定量刻画复杂网络?通过实证分析,了解实际网络结构的特点,并建立相应概念以刻画网络结构特征;二、网络是如何发展成现在这种结构的?建立网络演化模型,理解网络结构的产生和涌现;三、网络特定结构的后果是什么?利用基于网络的特定动力学过程,如新陈代谢网络上的物质流、食物链网络上的能量流、万维网上的信息传播、社会网络上的舆论形成等等,探讨网络结构与功能的关系;四、利用网络结构的设计和调整,控制和优化系统功能。

意大利的圭多·卡尔达雷利教授,是网络科学领域的知名专家,在复杂网络研究中,特别是社会、经济网络的分析和应用中颇有建树。《网络》一书,结合许多生动有趣的案例,如六度分隔实验、空手道俱乐部、万维网、食物链网络、基因调控网络等,深入浅出地介绍了网络科学的发展历程、核心概念和最新进展。书中没有一个数学公式,但通过具体案例的解读,能够让大家科学准确地把握网络分析的基本概念,如小世界性质、无标度网络的异质性、网络中的社团结构、无标度网络的脆弱性和鲁棒性、网络结构与传播的关系等,展现了作者强大的学术功力。可以说,《网络》一书,是大家了解和进入网络科学领域的非常好的一本入门读物。

总的来说,复杂系统的涌现现象是具有整体性和全局性的行为,不能通过分析还原的方法去研究,必须考虑个体之间的关联和相互作用。从这个意义上讲,理解复杂系统的行为应该从理解系统相互作用的网络结构开始。2009年,《科学》杂志以"复杂系统与网络"(Complex Systems and Networks)为主题,发表一集专刊,其中巴拉巴西教授的一段话很有启发意义。他指出,由于底层结构对于系统行为有着重大的影响,除非探讨网络结构,否则没有办法去理解复杂系统。希望《网络》一书能够带领大家进入生动有趣的网络科学领域。

献给我的家人
——圭多·卡尔达雷利

献给安娜
——米凯莱·卡坦扎罗

目 录

第一章　从网络的观点看世界　1

第二章　富有成效的方法　7

第三章　网络世界　22

第四章　连接与闭合　41

第五章　超级连接器　54

第六章　网络的涌现　66

第七章　深入挖掘网络　80

第八章　网络中的完美风暴　94

第九章　整个世界是否就是一张网？　110

　　　索引　115

　　　英文原文　125

第一章

从网络的观点看世界

网络存在于许多人的日常生活之中。通常的一天里,我们都会查收电子邮件,更新社交网络上的个人资料,打移动电话,使用公共交通工具,乘坐飞机,转移货币和货物,或者开始新的私人和业务关系等等。我们在所有这些情况中——有意或无意地——都会涉及网络及其特性。同样地,网络也会出现在重要的全球现象中。金融危机会在银行和公司的联系网络中产生多米诺效应。流行病,比如禽流感、非典型肺炎或者猪流感,则在机场网络之间蔓延。气候变化能够改变生态系统内部不同物种之间的关系网络,恐怖主义和战争则瞄准了一国之中的基础设施网络。电网内会发生大规模停电,计算机病毒在互联网中传播,政府和企业则会通过人们的社交网络和其他数字通信工具来追踪他们的身份信息。最后,遗传学的各种应用取决于人们对在细胞内部起作用的基因调控网络知识的掌握。

在所有这些情况中,我们都会处理大量不同元素(个人、公司、机场、物种、发电站、计算机、基因……)的集合,它们通过无序的多种相互作用模式而彼此关联,即它们都有着某种潜在的网络结构。通常,这种隐含的网络结构是理解前述重要现象的关键所在。其中一个很好的例子便是上世纪80年代大西洋西北海域鳕鱼种群的衰竭。当时,鳕鱼的短缺给加拿大水产业造成了巨大的经济危机。加拿大境内的利益相关者呼求更多地猎捕海豹,他们认为控制鳕鱼捕食者的数量有助于阻止鳕鱼种群的衰竭。大量的海豹在上世纪90年代遭到猎杀,但鳕鱼种群并未恢复。与此同时,生态学家们则对关联鳕鱼和海豹的不同食物链进行了研究。到上世纪90年代末期,他们已经为研究发现的关联这两个物种的大量不同食物链条绘制了一幅详尽的网络图(图1)。根据这个复杂的网络图,捕杀鳕鱼捕食者并不必然有助于鳕鱼种群的恢复。比如,海豹捕食大约150种猎物,其中也包括鳕鱼的其他捕食者。因此,海豹种群数量的减少便增加了其他鳕鱼捕食者对其种群数量所造成的压力。

生态系统乃复杂的物种网络:如果我们想要理解并管理这些物种,将这种底层的网络结构纳入考虑范围则至关重要。我们对其他一些基于网络结构的系统也必须采取类似的谨慎态度。例如,像艾滋病之类的传染病的蔓延就受到某些人群内部无保护性关系模式的强烈影响。同样,流动性冲击则取决于银行间货币交换网络的相互交织。

上述所有例子都是所谓**涌现现象**的例子。此种现象乃无法通过观察构成系统的单个元素而进行预测的集体行为。通常,呈现这些现象的系统被称为**复杂系统**。比如,单个的蚂蚁是相对笨拙的昆虫,但众多蚂蚁一起则能够进行如修建蚁冢或者储存大量食物这样复杂的活动。而在人类社会中,社会秩序产生

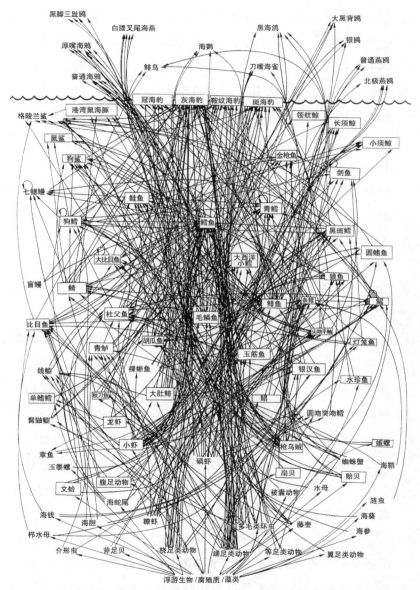

图1 加拿大东部西北大西洋"斯科舍大陆架"食物网局部图。箭头方向从捕食者指向猎物

于自主个体的结合,尽管他们的利益往往相互冲突,但最终仍然合力执行凭一己之力无法完成的任务。类似地,某个鲜活有机体的生命力来自其不同组分的相互作用;互联网应对错误、攻击和信息量高峰的非凡弹性是网络作为整体的性能,而非单个机器运行的结果。

网络及其所强调的互动,乃理解诸多此类现象的关键。设想两支足球队的球员水准十分相近,最终表现却差异甚远。这种差异很可能取决于球员们在场上相互配合的好坏程度。类似地,单个球员可能在联盟球队中表现尚好,而在国家队中却表现欠佳,因为该球员在两个团队中的相对位置有所不同。团队的表现是某种涌现现象,它不仅取决于单个参与者的才能或个体技艺的总和,而且还取决于参与者之间所形成的互动网络。许多涌现现象都严重依赖于其底层的网络结构。

网络方法将所有注意力都集中在了单个系统内相互作用的全局结构之上,而每个组成部分的具体属性均被忽略。结果,像计算机网络、生态系统或者社会群体等十分不同的系统都用同一种工具来描绘,那就是**图**(graph),即由多种关联所结成的一种单一的节点架构。这个方法最初由莱昂哈德·欧拉在数学领域发展起来,之后扩展至广泛的学科领域,包括大量使用这种方法的社会学,以及最近的物理学、工程学、计算机科学、生物学和其他诸多学科领域。

使用同一种工具来表现差异显著的不同系统只能以某种高度抽象的方式完成。系统细节描述的缺失在**普适**形式中得到弥补,即将十分不同的系统视作相同理论结构的不同实现形式而加以考虑。由此观之,计算机病毒的传播可能与流感类似;劫持一个路由器可能与某个物种在生态系统中的灭绝有着相同的影响；而万维网的增长则能与科学文献的增长放在一起比较。

这条推理线索提供了诸多洞见。比如，把系统呈现为图能让我们察觉到那些将明显不相关元素囊括在内的大规模结构。2003年，瑞士电网中的一个小事故触发了一场大规模停电事件，其影响波及1000公里之外的西西里岛。专注网络结构让我们看到，遥远事物之间最终可能通过难以置信的关系或传播短路径而强烈地关联着。地理和社会上相互隔离的两个个体——比如某个热带雨林中的居民和伦敦金融城中的某位经理——仅通过六度分隔（six degrees of separation）就能相互关联，这个最近的观察与现实相去不远。并且，这种现象可用社会关系的网络结构加以解释。

网络方法还揭示出系统的另一个重要特征，即某些不受外部控制而发展起来的系统仍然能够随机发展出某种内部秩序。细胞或生态系统并非"设计"而成，但仍以某种牢固的方式起作用。类似地，社会群体及社会趋势源于大量不同的压力和动机，但仍然展现出清晰和确定的图景。而互联网和万维网则在缺乏任何监管机构的情况下兴盛起来，并由大量不相关的因素所推动，但它们通常还是按照一致且有效的方式运转。所有这些现象都是**自组织过程**，即系统内部秩序和组织并非外部干预或整体设计的结果，而是局部机制或倾向在成千上万次相互作用中迭代产生的现象。网络模型能够以清楚和自然的方式描述自组织在诸多系统中是如何产生的。同样，网络让我们能够更好地理解诸如计算机病毒的快速传播、大规模流行病、基础设施的突然崩溃，以及社交恐惧症或音乐流行之爆发等各种动态过程。

在复杂、涌现以及自组织系统的研究领域（即关于复杂性的现代科学）中，网络作为通用的数学框架而变得愈发重要，涉及大量数据时尤其如此。这通常体现于个人在搜索引擎中的累积查询、社交网站的更新、在线支付、信用卡数据、金融交易、移动电话中的GPS（全球定位系统）位置等情况中。在所有这些情形下，网络对于整理和组织这些数据，进而将个人、产品和新闻等相互关联而言是重要的工具。类似地，分子生物学则越来越依赖计算策略以便在其自身产生的大量数据中找到秩序。科学、技术、健康、环境和社会等许多其他领域也是如此。在所有这些情况中，网络正成为揭示复杂性之隐藏结构的典范。

第二章

富有成效的方法

穿越欧拉之桥

在俄罗斯的加里宁格勒市,一座名为奈佛夫的岛屿坐落于普雷格尔河内。该市300年前曾隶属于普鲁士王国,彼时唤作哥尼斯堡,而奈佛夫岛与该市其余部分则由七座桥相互连接(图2上)。当时的城里流行着一个谜题:是否可能在不重复走过一座桥的情况下,遍历所有七座桥?之前没有任何人成功做到这一点。另一方面,对这种走法的可行性,当时也不存在任何形式证明。而其解决方案来自古往今来最著名的数学家之一。1736年,莱昂哈德·欧拉以一种不寻常的方式画出了哥尼斯堡市的地图。他将陆地和岛屿的部分表示为点,而桥则为连接各点的线(图2下)。

当我们以这种方式看待该问题,事情就变得更加容易了。通过展示城市的网络结构,欧拉证明了谜题中的走法不可能。他的解释基于以下观察:若此种走法可行,则路程沿线所有点的

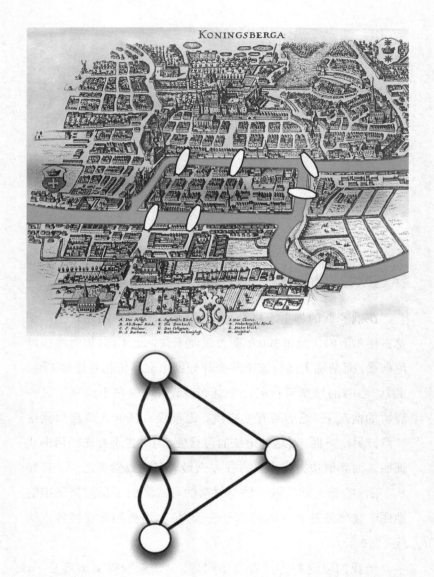

图2 版画中的哥尼斯堡（上），即现在的加里宁格勒，它表现了哥尼斯堡七桥之谜，莱昂哈德·欧拉将其表示为一幅简图（下）

连接数必为偶数。这是因为每当某人通过一座桥进入该市的任何部分时,他必须通过另外一座不同的桥离开。总之,每个区域必须有偶数座桥,比如2座、4座、6座等等。只有起点和终点可以有奇数个连接数:起点只能有一座桥,终点亦复如此。不幸的是,哥尼斯堡地图所有顶点的连接数均为奇数。因此,人们不可能一次性遍历所有桥而不重复。

哥尼斯堡地图这种简单的数学图是**图**的首个例子。数学家将构成这幅图的点和线分别称为**顶点**和**边**。如今,欧拉已经因为开启了一整个建立在图解分析基础之上的数学分支而被人们铭记。他的这种直觉可被认为是网络科学的首个创立契机。在他之后,许多数学家都曾研究过网络的形式特征,科学家则将它们应用于更加广泛的问题之中:1845年G.R.基尔霍夫的电路,1857年A.凯利的有机成分异构体,1858年由W.R.哈密顿提出的"哈密顿圈",等等。其中一个著名的应用是于19世纪中期提出的"地图着色问题"。当时,地理学家正在设法计算出绘制地图所需的最小颜色数量,其中相邻的国家应有不同的颜色。这不仅仅是个理论问题:考虑到为数众多的国家和印刷行业中数量有限的不同油墨,使用最少数量的颜色显得至关重要。从经验角度看,三种颜色不够用,四种颜色则似乎很奏效。人们直到1976年才证实了解决方案确为四种颜色。证明基于将一张地图以图的方式呈现出来,其中的节点为国家,边则画在两个共享一个边界的国家之间。

出走的女孩,澳大利亚人和芝加哥工人

1932年的秋天,在两个星期内有14名女孩从位于纽约州的哈德森女子学校中出走。这种出走率不同寻常。学校管理

9 层于是决定调查这些女孩的个性,以便理解这种现象。由于没有发现女孩们的性格有特殊差异的明显证据,精神科医生雅各布·莫雷诺便提出了一种完全不同的解释。他指出这种大规模的出走情形是由出走女孩们在社交网络中的位置引发的。莫雷诺和海伦·詹宁斯一起利用**社会计量学**,即一种能够识别个体之间关系的技术,绘制了学生之间的社会关系图。他们发现,这些关系是出走的想法在女孩们之间传播的主要渠道。个人在友谊网络中的位置对其模仿出走女孩的行为至关重要。

莫雷诺是第一批将网络理念应用于社会的研究者之一。继欧拉的直觉之后,他的研究成为网络科学创立的第二个关键步骤,它开启了网络科学最重要的分支之一:社交网络分析。30年之后,人类学家采用了类似的方法研究部落的亲属关系,比如对澳大利亚阿伦达人部落的研究。在这个案例中,互为亲属的人们之间的联系图得到绘制。研究人员发现,他们得到的结果图对应着优雅的数学结构。这些以及其他结果都表明,在人类社会的无序之下,暗含着精巧的社会结构,甚至是普遍法则。从那时起,社会科学便广泛运用网络概念来呈现社会结构。许多其他的研究也遵循着这些最初的研究方法,比如,对南美洲妇女群体(A.戴维斯、B.B.加德纳以及M.R.加德纳等人,1941)、芝加哥工厂的工人群体(E.梅奥,1939)、学童之间的友谊(A.拉波波特,1961),以及对科罗拉多州斯普林斯市吸毒者之间关系的研究(理查德·B.罗滕伯格等人,1995),诸如此类。

随机联系

网络科学创立的第三个重要时刻则随着保罗·埃尔德什和

阿尔弗雷德·雷尼这两位数学家在1959年到1961年间一系列论文的发表而到来。埃尔德什乃20世纪最重要的数学家之一，人们称他为"数字情种"。这句话不全对——他也钟爱着图形。这两位理论家研究了某种表示图的数学模型，其中的顶点完全随机地连接。数学研究人员雷·萨洛莫诺夫和阿纳托尔·拉波波特最早于1951年在一篇论文中提出了这种后来被称作**随机图**的模型。

随机图是一种十分简化的模型，其特征与那些真实网络截然不同。比如，随机性和概率的确可能在结交新朋友的过程中发挥重要作用，但是友谊网络的形成肯定也与许多其他因素相关，比如社会阶层、共同语言、吸引力等等。然而，随机图模型仍然非常重要，因为它量化了完全随机网络的性质。随机图能用作任何真实网络的基准或**无效状态**。这意味着，我们能将随机图与真实世界的网络进行对比，进而了解概率在多大程度上塑造了后者，以及其他标准在何种程度上起了作用。

以下是构建随机图的最简单方法。我们取所有可能的顶点对，然后为每一对顶点抛掷硬币：如果结果是正面朝上，则画出一个连线；否则我们转向下一对顶点，直到遍历所有顶点对（这意味着我们以 $p=1/2$ 的概率绘制顶点连线，但可能会用到 p 的任意值）。通常，图的创建及相关研究都不是人们手动完成的。科学家运用计算机程序，在图纸或电脑屏幕上绘制生成网络图。然而，当网络规模很大时，这项工作就会变得越发困难。此外，人们仅通过目测很难研究相互交织的结构。运用数学工具将图作为抽象对象进行研究能够提供更好的量化见解。计算机对此也有帮助：模拟使我们能够在计算机的"头脑"中将这种模型可靠地具象化，然后对其进行测量，就像它是个真实的物体一样。如果我们想要将抽象模型与真实世界网络进行比较，现在只需要比较两种情况下的测量数据。

自上世纪60年代被引进以来，随机图模型已成为最成功的数学模型之一，尽管它与现实的联系并不紧密。现在，它是所有网络进行比对的基准，因为任何对此模型的偏离都表明了在许多现实世界网络中存在着某种结构、秩序、规律性和非随机性。

捕获信息之网

当纽约、马德里、伦敦分别于2001、2004、2005年遭遇死灰复燃般的恐怖袭击之后，一些政府机构提议存储电子信息流数据，并将其作为一项反恐措施。在这项提议中，公民之间多年的电话和电邮都会以安全的名义被记录在案。具体的通信内容则不会被储存，仅记录信息的发出者和接收者便已足够（有时候还包括通信的时间和地点信息）。正如警方所知晓的那样，即便这种简单的谁与谁之间相互关联的图景也是追踪人们活动的强大工具。的确，略略看一下某人的电话记录，我们就可以推断出他的习惯、朋友圈以及各种相关的其他数据。

这是网络科学基本方法的一个非常实际也颇具争议的例子。复杂系统由图——通过同样的相互作用而彼此关联的一组等价元素——表示，却忽略了其组成部分的详细特征，以及它们之间关系的具体性质。这一做法似乎过于激进，但它仍然让我们能够捕获到比乍看之下可预期到的更多的信息。这种方法的一个有效证明便是内嵌于诸多在线社交网络中的朋友推荐系统，比如"脸书"或"领英"。其中的想法很简单：你很可能认识你朋友的朋友。虽然很简单,却在多数情况下都有效。这种方法被称为**有根据的推测**,它也是很多在线商店的书籍或其他商品推荐系统的基础。这些公司的软件利用了与每个消费者相关的商品网络。这就是为何商业公司会储存大量的电子数据,包括电子邮件和在线社交网络数据——它们知道这种信息十分有用。

网络科学的基本方法可应用于广泛的系统之中。例如,一个由数百个物种组成的生态系统,其中每个物种通过不同的捕食策略从其他物种中提取能量。在网络方法中,这种情况由一组相同的边连接起来的一组相同顶点表示。同样的策略被应用于从互联网(通过电话线、光纤电缆、卫星通信等方式连接起来的无数计算机、路由器和交互站点等)到人口族群(通过各种关系相互联系且拥有不同目的和身份的大量行动者)的各种系统之中。尽管网络方法消除了所考虑现象的许多个别特征,它仍然保留了个体的某些特征。也就是说,这种方法并不改变系统的大小——系统元素的数量,也不会改变相互作用的模式——元素之间的特定连接集合。然而这种简化的模型仍然足以捕获系统的特性。

从个体到群体

在处理多种不同元素以不同方式相互作用的情况时,存在两种可能的方法。第一种方法是确认这种情况中的基本成分以及它们之间的相互作用。通过研究每个元素本身,我们可以将系统的行为推导为个体元素的加总。例如,生态学家可以通过列出每个物种的猎物和捕食者来描述生态系统的特征。计算机科学家通过关注每个不同机器的特征和协议来描述计算机网络。心理学家则通过描述每个社交者在他的圈子中的行为来研究社会关系。与第一种策略不同,第二种策略旨在将许多元素组合成为若干同类的群体。例如,社会学家和政治学家通常按照社会阶层、性别、教育水平、种族、国家等因素对社会进行划分。同样,流行病学家则常常将人口分成一组有限的"区划":健康的、感染的、免疫的个体等。生态学家也能用同样的方法将所有在食物网中扮演类似角色的物种汇集成群组(即**营养物种**)。

网络方法试图补充这两种策略。若我们只关注个体元素的行为,许多现象就无法得到解释。例如,生态系统内物种的数量动态并不单单依赖于该物种自身的特征,猎物–捕食者相互作用的全局网络也必须考虑在内。另一方面,专注于大类的元素也可能没什么用。比如,一国发生的政治现象基本上不是由先前存在的国家认同引起的,而是该国内部社会关系的具体模式运作的结果。网络方法则介于通过单个元素和通过大群组进行的描述之间,并将二者连接起来。在某种意义上,网络试图解释一组孤立的元素是如何通过相互作用的模式转化为群体和共同体的。在与这种模式相关的所有情况下,网络方法都提供了重要的洞见。

地理与"网络图"

20世纪初,伦敦的地铁服务变得十分复杂,以至于人们必须不时地发布愈来愈多的地图以便为出行者提供向导。1931年,经过多次尝试,地铁公司的雇员亨利·贝克改变了绘制地铁路线图的标准。与那种将地铁路线嵌入伦敦实际地图之上的做法相反,贝克将它们置于抽象的空间中(图3)。站点由间

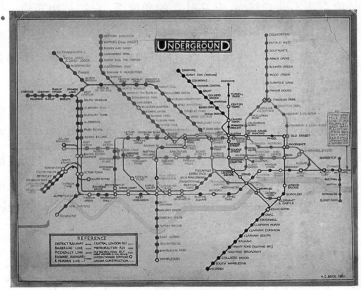

图3 伦敦地铁的"度量"示意图(上)与"拓扑"示意图(下)。尽管车站的实际位置和相对距离都未呈现在拓扑图中,它却是地铁服务的更佳思维导图

隔合适的点表示，而线路的连接则全部变成了具有45或90度角的直线。这个地图无关车站的真实位置及其相互之间的实际距离，但它对乘客而言更加清楚且有用。乘用地铁网络的人对其地理特征不感兴趣，有车站的顺序和地铁路线的交汇信息便已足够。

亨利·贝克的伦敦地铁图基本上就是一个图。他解决映射问题的方案利用了网络方法的一个基本特征：在网络中，**拓扑**比**度量**更为重要。也就是说，何物与何物相互关联比它们相隔多远更加重要。换句话说，图的实际地理情形不如其"网络图"重要。这两个概念的差异如图4所示。从度量角度看图中呈现的三个图形有所不同。也就是说，空间中的节点位置和节点连线的长度有所不同。然而，从拓扑学的角度看，它们是相同的——它们只是同一个图的三种不同呈现方式。在网络的表示方式中，系统元素之间的关联比它们的相对距离以及在空间中的具体位置要重要得多。

对拓扑学的关注是网络方法的最大优势之一，当拓扑与度量相比更为适用时，网络方法就是有用的。例如，从纽约发送至伦敦某办公室的电子邮件会与来自其隔壁办公室的邮件同时抵达。即便在互联网这种嵌入地理空间的实体基础设施中，连接的模式也比物理距离更加重要。在社交网络中，拓扑的实用性

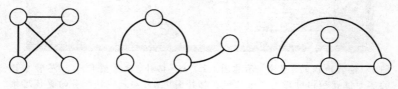

图4 同一个图的三种不同表示

意味着**社会结构**很重要。里奥内尔·梅西是当今世界最好的足球运动员之一。然而，他的表现却随着他所效力球队（比如阿根廷国家队或者巴塞罗那足球俱乐部队）的不同而有所不同。一些社会科学家认为，梅西在阿根廷国家队中与其他队员之间的关系网络不同于他在巴塞罗那足球俱乐部队中的关系网络。根据他们的研究，这会导致球员在前一种情况下背负更重的"负担"，这至少能部分解释他在不同球队中的表现差异。类似的现象也出现在更加复杂的社交"游戏"中，个人在其中的境况会受到他或她在关系网络中所处位置的强烈影响。

链条、网格和网络

网络方法将复杂系统简化为单一的点线架构。这是一个重大的简化，但其得出的结果图却可能不是那么容易得到解释：这便是图4所示的棘手例子展示的情况。甚至一幅仅由节点连接而成的链条图也可以是一个处理起来相当复杂的对象。一个链条可以代表比如搬运一桶水的消防队；或者物种的食物链，其中，第一个捕食第二个，第二个捕食第三个，以此类推；或者企业之间的供应结构——一系列公司中每个公司供给下一个公司。

想象一下五家公司（公司1, 2, 3, 4和5）的生产链。在这个链条沿线，其中任何一家公司都能与其两个邻居之一达成交易。其中的规则为，每个公司只能签订一份合同——例如，如果公司3与公司2达成了交易，则它不能与公司4另有协议安排。给定了这个简单的结构和规则，结果便是位于节点1和节点5的公司议价能力更低，因为它们的选择更少。这导致公司2和公司4更加强大，并且（意外地）削弱了位于节点3上的公司。确实，

17 节点3上的公司只得应付更加强大的公司，因此，它最终只能达成不那么有利的交易。简单如节点线性序列这样的事物确实产生了相当复杂的景观。社会学家将这个例子呈现的现象称为**排斥机制**。这远非理论中的情况，而是经济学中经常能遇到的情形——当两个公司之间建立商业关系以排除第三方时便是这样。

　　为了进一步深入这个问题，人们必须考虑到现实中的系统绝少如链条这般简单。在过去，科学家用规则网格或**晶格**表示复杂的系统，而非使用图。这些对象由许多元素组成——分别代表人、动物、计算机等等——它们通常按照规则的连接模式排列，例如棋盘上的棋子仅与其四邻的棋子相连。相对于用图表示而言，这种规则结构使得系统能够更容易被数学计算和计算机模拟这两种方式处理。

　　尽管格子相对于图而言更加简单，但它引入了强烈的限制条件。事实上，一个格子仅适于呈现经过精心设计或受到强烈约束的系统。比如，这些系统可能是某个计算集群的处理器阵列，或者某个嘈杂工作场所中彼此关联的工人之间的语言交流。在格子框架中，每个节点都连接着固定数量的最近连接点，而在绝大多数真实情况中，这些连接则关联着数量可变的元素，无论它们彼此是邻近还是相隔很远。捕获这种混乱状态的能力是网络方法的最大优点之一。

　　大部分这种无序状态都被编码为某种关键的数量——**度数**，即附着于每个节点之上的边数。如果节点是网页，度数则表示了该网页与其他页面的链接数量。如果节点是物种，度数则

18 表示了该物种赖以生存的物种数量。如果节点是一个人，度数

网络

则表示了其熟人的数量。这个圈子能与社会学家彼得·马斯登所谓的**核心讨论网络**联系起来，即那些可与之讨论重要事宜或打发时间的人群（朋友、伙伴、家庭成员、现在和过去的同学、同事、邻居、俱乐部伙伴、专业顾问、咨询员等等）。

描绘关系

两个人之间可以有无限的可能关系。他们可能共有同样的态度、想法或性别，可能是朋友、亲戚或同事，也可能是性伴侣或者仅仅效力于同一个足球队。此外，这些关系中的两个或更多可以同时发生在同样两个人身上。其中一些是合作关系，另一些则表示公开的敌对，而介于两者之间的关系则有着广泛的可能。最后，一些关系可能仅仅单方面被感知到，而另一方则完全对其忽略（比如，摇滚明星的粉丝可能感觉自己与其有某种联系，而摇滚明星可能完全忽略了他们）。社会学将个体之间广泛的可能联系进行了分类（表1）。在社交网络中具有多种关系类型的倾向又被称为**关系的多重性**。但这种现象也出现在许多其他的网络中——例如，两个物种可以通过不同的捕食策略而彼此关联，两个计算机则通过不同的有线或无线连接彼此关联等。

我们可以修改某种基本图来考虑关系的多重性，例如，为关系连线贴上特定的标签等。比如说，我们可以将某种关联是正面还是负面纳入考虑范围。物种通过捕食（负面）或共生（比如在开花植物和授粉者之间建立起的正面关系）而彼此关联。人们可以是敌人（负面）或朋友（正面）。一个网页则可以连接到另外一个网页以批评其内容（负面）或为其做广告（正面）。通

表1 社交网络关系的分类

相似点			社交关系				互动	流动物
位置	成员	属性	亲属关系	其他角色	情感关系	认知关系		
例如：处于同一个时间和空间	例如：同属一个俱乐部经历同样的事情	例如：同样的性别同样的观点	例如：母亲兄弟姐妹	例如：朋友老板学生竞争者	例如：喜欢憎恶	例如：了解知道认为……幸福	例如：与某人有性关系与某人交谈过给某人建议帮助某人伤害某人	例如：信息信念人事资源

过增添这个简单的二进制功能便让事情极大地复杂化了。想象一个由爱丽丝、鲍勃和卡罗尔组成的三人群体。当他们所有人由正面的关系联系时，一切都安好。又或者，爱丽丝和鲍勃可能是朋友关系，但他们二人均对卡罗尔抱有敌意。当情况相反的时候，事情就变得复杂了：爱丽丝与鲍勃和卡罗尔二人都保持着正面的关系，但鲍勃与卡罗尔之间却彼此憎恶。而当他们每个人都彼此憎恶的时候，情况就变得着实棘手了。根据社会学理论，第一种情况和第二种情况**在结构上是平衡的**，而第三种和第四种情况则相反。2006年，数学家蒂博尔·安塔尔和物理学家保罗·克拉皮夫斯基、西德尼·雷德纳将这个概念应用于第一次世界大战前六个欧洲国家之间外交联盟关系转变的分析之中。他们证明了这六国的联盟逐渐演化成了一种结构上平衡的状态，在这种情况下，强大的联盟得以建立，或者明确的共同敌人得以确定。这六国后来分裂为两个集团（一方为英国、法国和

俄罗斯；另一方则为奥匈帝国、德国和意大利），每一方的成员国彼此在集团内相互结盟，而对另一方的所有国家保持敌对关系。这种情况出现后不久，战争便爆发了。这个例子表明，结构上的平衡并不必然就是有利的。

图论让我们能为边编码更加复杂的关系，当关系并非对称的时候更是如此。狼捕食羊，博客链接到大报，一些人爱上另一些人；相反的情况绝少为真。在这种情况下，图中的关系乃某种单行道，我们可在其中朝着一个方向前进，但不能逆行。如果给边添加方向属性，则得出的结构为**有向图**，其中的关系由箭头指示。在这些网络中，我们有**入度**和**出度**分别测量节点向内与向外的连接数量。

到目前为止，我们所考虑的关系都是二进制的——它们仅有两个值。这种**二分**关系要么存在，要么不存在：例如，与某人结婚或者被某人雇用。然而，这种关系在统计上都是例外：绝大多数情况下，关系展现出广泛的强度变化。捕食以吃掉猎物的数量计算，网页则能被零星或大量的链接关联，而爱情的强度则从轻微的吸引到强烈的激情不等。这些进一步的特征对应于我们赋予关系的**权重**。例如，**加权网络**可能作为个体或事物之间不同交互频率的结果出现。

我们对基本图结构的其他修改也是可能的，且处理这些事物的技术非常有趣。比如，大部分社交网络研究致力于了解不同种类的关系如何互相影响。然而，网络方法的优点在于，在某些情况下，忽略所有或大多数具体细节是合理且有效的：有向网络变得无向，权重被移除，多重连接坍塌为某条单独的边等等。结果表明，这种彻底的简化仍然能够捕获大量信息。

第三章

网络世界

网络组学

上世纪80和90年代,人们倾向于认为在某种方式上一切都是由基因决定的。报纸上的报道都与"同性恋基因""肥胖基因""暴力基因"或者"酗酒基因"有关。这种态度呼应了人们对人类复杂性的秘密隐藏在基因组中的期望。DNA——细胞核中包含基因的脱氧核糖核酸分子——又称"生命的软件",该程序负责生命体的每一个特征正是其中代码功能的障碍导致了所有的疾病。这种图景引发了人们对基因组进行测序的风潮,而人类基因组图谱于2001年2月的发布将这个风潮推向了顶点。测序的结果令人十分惊讶,人类的基因数并不比线虫多多少,而且比某些种类的水稻基因还少。人类几乎与类人猿有着相同基因组的推断是合理的,但问题是人类基因组也与老鼠类似。软件的隐喻并不支持这一证据:DNA序列本身并不能解释我们所能观察到的物种差异,更不用说单个个体的所有特征和疾病了。事实上，生物的基因与其相应的宏观特征之间还隔着一系列漫长的步骤。此间的变异决定了不同的结果。

基因之上的第一层复杂性由**基因调控**给出。包含在DNA中的基因被转录和翻译以产生蛋白质。蛋白质几乎在生命的各个方面都起着核心作用：肌肉运动、血液循环、承担酶的作用、结合激素等等。此外，蛋白质还彼此相互作用：蛋白质的产生可能受到细胞内现有蛋白质的促进或阻碍。这些相互影响之间的微妙平衡对于生命而言至关重要。例如，p53这一单个蛋白质的突变就伴随着大量不同的癌症。这些交织的激活和抑制模式产生出了**基因调控网络**。在这个网络之中，节点为基因，连线则是关联某个基因与其他基因表达的反应链条。

蛋白质以所有可能发生的方式相互作用，这代表着另外一个层面的复杂性。例如，一些蛋白质可以结合在一起。这些大分子像分子机器一样起作用，它们在细胞这种机器中发挥着自己的功能。为此，蛋白质必须具备正确的几何形状以彼此契合。当蛋白质以错误的方式折叠时，就可能出现一些问题。比如，人体中对应"疯牛综合征"（又称克雅二氏病）的蛋白质（即**朊病毒**）就被认为是错误折叠的。蛋白质之间所有可能的实体联系可表示为网络。在**蛋白质相互作用网络**中，顶点为蛋白质，而如果蛋白质在细胞中的确相互作用，则可在相应的点之间画出一条边。

蛋白质并不足以让细胞工作。细胞通过许多不同的分子与外界环境交换着物质、能量和信息，参与其中的反应数达百万级。饥饿、饱腹感、寒冷以及生物体大致上所经历的所有状态都赖于这种被称为**新陈代谢**反应的集合。而通过一系列中间步骤将一个分子转化为另一个分子的反应链则被称为**代谢途径**。然而，细胞中的反应很少遵循有序序列的模式。例如，反应终端的分子经常会与开端处的分子相互作用以终止反应。这种反馈过程就闭合了反应链中的一环。所有这般路径的集合就产生了复杂的**代谢网络**。

因此，生物体是几层网络共同作用的结果，而不仅仅是简单基因序列的决定性结果。基因组学中已经加入了**表观基因组学**、**转录组学**、**蛋白质组学**、**代谢组学**等研究相关层级的学科，这通常被称作**组学革命**。而网络正是这场革命的核心。

思维之网

直到18世纪，"灵魂"可以体现在某个器官中都是一个奇怪的设想。但是，医生已经意识到中风或者其他脑损伤可能会危及关键的认知功能——心与脑之间的关联从那时开始变得明显起来。当时，解剖学家弗朗茨·约瑟夫·加尔便敢于提出所有心理官能必然起源于大脑这种想法。他确定了大脑中的27个"器官"，每个器官分别负责颜色、声音、记忆、言语，以及友谊、仁慈、骄傲等等。这种想法听起来如此异类，以至于加尔不得不逃离维也纳，进而在激进的法国找寻避难所。

后来，几位生理学家试图通过比如从鸽子的大脑中去除切片这样的方式验证加尔的理论。然而，他们并未找到加尔所设想的器官存在的任何证据。因此，他们得出结论说，大脑是产生思想的一个均匀、未经分化的统一体：如其中一位生理学家所说，"大脑分泌思想就像肝脏分泌胆汁一样"。这种观念在1860

年代保罗·布罗卡的研究出来之前一直占据主导地位。在表达性失语症患者的尸体解剖中,布罗卡总能在他们大脑左侧的额叶区域发现一些损伤。在确定了人们现在称其为**布罗卡区域**的大脑部分之后,他便宣称,"我们用左脑说话"。从那时起,神经学家找到了负责不同活动的多个功能中心,但他们也发现这些中心区域绝少孤立运转:大脑不同区域的整合对其功能而言至关重要。

网络在被划分为专门领域的大脑模式和作为整体的大脑模式之间架起了桥梁(这与社会科学的情况并无不同,在这个领域中,网络允许我们从个人和共同体之间的某个层面描述社会)。大脑中遍布网络,其中的各种网状结构在专门的区域之间提供了整合。在小脑中,神经元形成了一个个不断重复的模块:它们的相互作用被限制在了相邻的模块之间,这与晶格中发生的事情类似。而在大脑的其他区域,我们发现了随机连接现象,即局部、居间或者遥远神经元之间的连接概率大致相等。最后,新皮质——哺乳动物的许多高级功能所涉及的区域——则将局部结构与更加随机且遥远的连接相结合。一些科学家认为这些线路体系可能负责主体意识:涌现的良知便可能是某种足够复杂的网络结构作用的结果。

确定这些神经网络的实际结构十分困难,因为细胞数量巨大,并且探测它们困难重重。我们仅掌握了那些十分简单的生物的详细神经网络图,比如一种名为秀丽隐杆线虫的寄生线虫。这种一毫米长的透明生物能存活三周,它仅有300个神经元,却是分子生物学界的超级明星。秀丽隐杆线虫是一种**模式生物**,即特别适于实验的生物,因为科学家对其特征了如指掌,且它在某些方面与人体类似。这种半透明的蠕虫通常是新药和新疗法实验的第一个基准。

目前，人们还不可能为人类大脑绘制类似的神经网络。然而，我们可以使用另外一种策略。当人们执行一个动作，即便简单如眨眼，来自神经元的电信号风暴就会在大脑中的数个区域爆发。这些区域可以通过诸如**功能性磁共振**这样的技术加以确认。通过这种技术，科学家发现不同的区域会发出相互关联的信号。也就是说，它们显示出某种特殊的同步现象，这表明它们可能相互影响。可将这些区域视为节点，而如果它们之间存在足够的相关性，则它们两两之间便可画上一条边。同样在这个层次上，大脑表现为相互连接的元素集合。人的每个动作都会点亮脑中的某个连接区域。

盖亚的血管

1999年，旧金山湾区经历了大规模的藻类暴发。通常，藻类的这种暴发是农业密集用地的结果：我们将氮、磷等化肥排入海中后，它们就会成为藻类的营养物。然而，本案例中的情况并非如此，因为一些政策和限制已经减少了从不同河流排入海中的营养物污染水平。加利福尼亚州的生态学家们通过搜集三十年的观察数据得出结论说，藻类暴发有着复杂得多的原因。1997年和1998年，人们记录到了最强的"厄尔尼诺"现象之一，紧随其后的便是1999年同样强大的"拉尼娜"现象。这些现象导致了当时加利福尼亚生态系统的变化。深海处冰冷的富营养化海水涌至海岸。这些营养物质吸引了海洋中的生物——鲽鱼和甲壳类动物——进入湾区。这些动物是湾区双壳类动物的捕食者,而后者反而是藻类蔓延的障碍。双壳类动物由于捕食者的增加而种群崩溃,这成为藻类暴发的直接原因。触发这种多米诺效应的条件可能来自正常的气候波动。然而,其后果对人类而言却是一种警告:气候变化,特别是极端气候的频率增加,可能对生态系统产生意想不到的影响。

  旧金山湾区海藻暴发背后的核心结构是**食物链**——一系列物种的连接:鲽鱼和甲壳类动物捕食双壳类动物,双壳类动物消耗藻类。生物以食物链的方式从彼此身上提取他们生存所需的能量和物质(这并非生态系统内物种之间唯一可能的相互作用:生物体也可以建立互利的相互作用,比如开花植物与其授粉昆虫之间的相互作用)。每个食物链都从**基位**物种开始,比如植物和细菌。这些物种不会捕食其他物种,它们通过转化阳光、矿物质和水而直接从环境中获取资源。这些资源以接连的捕食方式沿着食物链顺次转移。**中位**物种既是捕食者也是猎物。而**顶位**物种(处于食物链顶端)则不被任何别的物种捕食。食物链帮助我们理解为何渔业会崩溃,比如70年代秘鲁鲥鱼产业的境况。在大规模的无差别捕鱼期之后,像鳕鱼或金枪鱼等捕食者就会大幅减少。在这个阶段之后,捕鱼业则朝着更加基位的物种进发,比如鲥鱼。但这些物种也会迅速崩溃。原因在于,当大型捕食者被移出食物链之后,食物网下游的其他捕食者便会取而代之。后者往往是不可食用的鱼类:它们的种群没了限制之后,便会耗尽其他可食用的基位物种。

  生态系统的实际情况甚至更为复杂:通常,食物链并不孤立存在,而是以复杂的模式彼此交织,其中的某个物种会同时属于多个食物链。例如,一个特化物种可能只捕食一种猎物(或者在

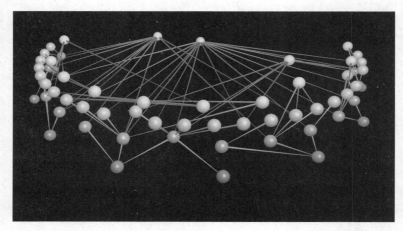

图 5 英国某草原食物网：节点代表英格兰和威尔士一些草地中的物种，连线则由捕食者（较粗的一端）指向猎物（较细的一端）

一些情况下仅捕食少数物种）。如果猎物灭绝，这个特定物种的种群就会崩溃，并导致一系列的**共同灭绝**。更加复杂的情况是，杂食动物捕食某种食草动物，而两者都吃某种植物。杂食动物种群的减少并不意味着植物的繁茂，因为食草动物将从这种物种减少中受益，进而消耗更多的植物。

将更多的物种纳入考虑范围，种群动力机制会变得越发复杂。这便是为何对生态系统而言，**食物网**（图 5）是比"食物链"更为恰当的描述。这些网络中的节点是物种，连线表示它们之间的捕食关系。连线通常是有向的（大鱼吃小鱼，而非相反）。这些网络提供了物种之间的食物、能量和物质交换，并因此构成了生物圈的循环系统。它们构成了地球的血管。

人类"角斗士"（网络人）

口口相传（word of mouth）是获得有关白领职位空缺信息

的惯常方式。所以，如果我们正在寻找这种工作，在亲友之间传递相关信息则不啻为一个好主意。不那么明显的是，我们将信息告知远方不常见面的熟人甚至会有更好的效果。这便是马克·格兰诺维特于1973年提出的观点。这位社会学家采访了波士顿郊区职业人士这一样本，他们当时都依靠个人联系获得了各自的工作。格兰诺维特问他们在获得工作之前与提供工作的人的联系频率。多数人的答复是"偶尔"，许多人的回答是"绝少"。与亲密的朋友相比，工作机会更可能来自旧时的校友、过去的同事和以前的雇主。运气或共同的朋友则是再现这些联系的管道。格兰诺维特将这种现象描述为**弱连带优势**。

格兰诺维特通过描述一位假想中的名为埃戈的人的熟人圈来解释这种结果。埃戈每天都和家人及一些亲密的朋友生活在一起。所有这些人很可能也彼此联系紧密。其结果便是，信息会在这群人中快速传播。因此，埃戈可能知道这群人中的所有新鲜事。与此相反，较弱的人际联系也会将他与遥远的人联系起来。这些人并不会受到埃戈社交环境的限制。因此，他们向他开放了一组新的群体集合，每一群人都封装着别的群体无法获取的信息。弱联系的缺失会为组织、公司或机构带来各种困难。信息和技术被限制在一个群体之内，而没有抵达那些需要它们的人手中。因此，这些东西需被重新创造或从外部顾问处有偿获得。据报道，惠普的前首席执行官曾感叹道："惠普不能闭目塞听！"格兰诺维特的直觉后来发展成为**社会资本**理论。这种想法意味着一个人的联系人（以及这些联系人的联系人）使他或她能够获取那些最终可以提供更好工作和更快升迁的资源。更普遍的是，个人在他或她的社交网络中的位置对于决定其未来的机会、限制和成果而言至关重要。

　　衡量交情并不容易,因为它是十分主观的事情。通常,像公司流程图之类的图谱并不是很有用,因为它们并不对应着员工之间的实际关系。正因为如此,流程图并不能帮助我们了解公司内部的信息渠道(以及可能的瓶颈)。科学家们设计了大量的替代策略来绘制社交网络,从问卷调查到**雪球式抽样**,即系统中的每个受访者都推荐自己圈子里的某人为下一个采访对象。这些策略使得数据的搜集就像不同个体(从美国中西部的高中生到非洲布基纳法索的村民)所组成的群体之间的性行为图那般敏感:有关这些关系网络的知识能让人们更好地理解性传播疾病的蔓延。

　　另外一种相对容易定义的关系则是专业协作。这样的网络存在于多个领域,从好莱坞(如果两个演员出演同一部电影,他们便会产生联系)到科学领域(如果两位科学家共同撰写一篇论文,他们也会产生联系)。合作还能在更不寻常的环境中找到,比如政治(美国参议员会在支持共同法律的基础上产生联系)或者恐怖主义活动(恐怖分子则基于情报和法律文件相互联系)中。

　　信息技术为衡量人际互动提供了一种强大的新手段。两人之间频繁的电话和邮件往来,或者在脸书、领英等虚拟社交网络中的友谊等,都表明了一种稳定的关系和网络图中的一条边。越来越多的公司利用他们客户的社交网络来找寻这样的信息。例如,据报道,电话公司会以提供工作机会或别的策略来瞄准"有影响力"的个人:当这些客户改换公司的时候,就会在他们的密切联系人中触发类似的变化。

语词之网,思想之网

玛格莱特王后在莎翁的《亨利六世》三部曲的中篇(第一幕,第三场)里说道:"难道说,亨利王上老要在乖戾的葛罗斯特管辖之下当小学生吗?"[①]王后抱怨葛罗斯特公爵对其国王丈夫的影响。她用"小学生"(pupil)表达的是什么意思?分类词典表明,"pupil"还可以表述为scholar(学者)、acolyte(助手)、adherent(信徒)、convert(皈依者)、disciple(门徒)、epigone(追随者)、liege man(臣下)、partisan(党羽)、votarist(拥护者)或者votary(崇拜者)等等。这个清单为"一个人从属于另外一个人"这项语义提供了一幅全景语词图谱。我们可以通过探索与"pupil"相关的语词即其**语义区**来扩展这一图谱。其他的语词包括faithful(信徒)、loyalist(忠诚者)、advocate(倡导者)、backer(赞助者)、supporter(支持者)、satellite(随从)、yes-man(唯唯诺诺之人)……那么,王后意欲用何种意义指谓其夫君呢?"pupil"的反义词表示non-student(非学生)、coryphaeus(领导者)、leader(首领)、apostate(变节者)、defector(背叛者)、renegade(叛徒)、traitor(背信弃义之人)以及turncoat(叛贼)等。当然,她在要求亨利国王反对葛罗斯特公爵的权威。

这是语词之间如何相互联系的一个简单例子。事实上,严格的分析应该将语词使用的历史差异、莎士比亚作品中某个语词的特定出现、它在剧本中使用的语境,以及其他诸多方面都考虑进来。公平地讲,在任何情况下,我们都可以更好地理解语词

---

① 译文引自朱生豪先生译本。——译注

的含义，若我们将其在语言中的"邻居"考虑进来的话。**同义**、**反义**以及**语义联系**仅仅是语词间少数几种可能的联系。其他则包括**整体-部分关系**和**上下位关系**等（"啤酒"是"饮料"的一部分，后者则是前者的上位词）。"pupil"为语词间的另一种联系提供了清晰明了的例子。这个语词有着迥然不同的两种含义：它表示学生，也表示眼睛的一部分[①]。这便是**一词多义**的情况。自然，莎翁的戏剧语境当下就决定了其正确的含义。一般来说，上下文提供了语词的具体含义：句子中语词的**共现**定义了它们自身的含义。如此的共现提供了语词的另外一种关系。例如，语词"国王"和"亨利"在英语句子中就比"国王"和"相对论"一起出现的可能性要大得多。

我们现在可以创建自己的语言地图。我们用语词作为顶点，同义词、反义词和多义词（这些语词关系可按照分类词典或一般词典绘制，而共现的模式则可从大型语言数据库中获取，比如英国国家语料库）则由边来连接。语义联系更加难以确定：对它们的研究构成了一个完整的语言学领域。一些语言具备将一个语词与一组相关语词联系起来的特殊词典。另外一个替代方案便是词汇联想实验。一个语词被提供给样本中的一群人，要求他们说出听到这个语词之后想到的第一个语词。得出的语词接着又被用来重复这个联想实验。以这种方式，我们可以一步步建立自己的语词联系之网。语词网络的不同实例代表了不同的结果。这取决于语言本身、文本的类型、文本作者所受的教育，或许它与语言功能障碍也有关系。

---

① **"pupil"** 还有"瞳孔"的意思。——编注

语词网络包含大量信息，但通常，这些信息在研究文本的具体内容和不同文本所表达的思想之间的关系时并不是特别有用。这是一个关键问题，比如对网络查询来说，通常，人们必须引入复杂的算法来执行该任务。然而，在一些文本中，人们可以绘制十分精确的网络。科学文献中即是如此。生产知识从来不是单个人的努力。如哲学家沙特尔的伯纳德在12世纪首次指出的那样，科学家总是"站在巨人肩膀上的矮子"。科学家的工作几乎总是建立在先前的结果之上。研究者们认识到了这一点，他们会在其论文末尾引用一些较早的出版物。引用提供了对过往相关成果的认可，赋予新的成果以可信度，并参考那些在某项研究中被接纳为有效或被批判过的事实、技术和实验。近年来，出版物在很大程度上实现了标准化：文章大多以英文写成，控制方法已经同质化（主要以**同行评议**的方式），文章影响力的衡量标准也已制定等。同时，出版物的大型电子数据库业已建立，其中每天都会增加上千个新项目：文章、书籍、专利和工程等等。所有这些生成了一个大型的出版网络：如果一个项目引用另外一个，则它们相互关联。我们还可从这些数据库中识别作者身份，并创建科学家之间的协作网络。人们越来越多地利用这些系统绘制知识发展和科学最活跃领域的图景，并将其可视化。

互联的货币

2008年，美国的一些大型金融机构突然破产。几个月的时间里，大多数发达国家都卷入了这场史上最严重的金融危机之一。关于这场危机的原因已经有过很多分析了。可以肯定的是，它表明经济在全球层面的相互关联已十分紧密。

古典经济理论将经济参与者表示为独立、完全理性的行为者,他们专注于最大化自己的收入。然而,事实表明个人、公司、机构和国家并非相互独立:每一方都以多种方式彼此影响着。他们的行为远非完全理性,而是强烈地依赖于主观态度、情感和相互影响。

贷款是公司和机构能够紧密互联的方式之一。一个有趣的例子是私人银行间日常会进行货币交换,以便能够满足银行客户的可能要求(因此变得**更具流动性**)。如果客户的要求超过了某银行的流动性储备,该银行可以向其他银行借款。世界各地的中央银行会要求其他银行将其存款和债务的一部分置于它们那里,以创造一个应对流动性不足的缓冲器。在这个意义上,中央银行确保了银行系统的稳定性,从而避免了流动性冲击。同业拆借网络的冻结是2008年金融危机的最初信号之一。

与贷款相比,**持股**,即某公司资本直接加入到另一个公司中,能够带来甚至更加紧密的联系。这意味着一家公司持有第二家公司的一部分,并能对其主要决策施加影响。当某公司持有大部分股票,或当它能够决定董事会多数成员的投票时,持股就转变为控股。在这种情况下,法律上独立的公司便转化为商业集团。通常,这些集团展现了某种金字塔结构,其中**控股**公司的权力在顶部,运作公司则处于控制层级的底部。

在大多数国家,商业集团以具体形式存在,并在法律上受到监管;但是更软性和更少规制的影响形式也可能存在。这在董事会内部最为常见。管理者经常同时在多个公司的董事会任职。显然,他们扮演着董事会之间信息、联盟或者利益的交换渠道。他们在不同董事会中的同时任职为相关公司之间建立了**连

锁关系。如果这些公司是明确的竞争者关系,这种情况则明显与自由市场不兼容。共同的董事要么支持其中一个公司而打压另一个,要么在公司之间建立垄断联盟(这通常会被法律排除掉)。一般而言,这样的董事会发现自己身陷来自不同公司的投资者之间的利益纠缠之中而难以运作。

公司之间相互关联的进一步证据来自**股票价格的相关性**。财务人员知道在同一行业(例如采矿、运输、服务、食品等)经营的公司的股票会以某种类似或"同步"的方式改变价格。例如,电子行业(或任何其他行业)内所有公司的股票价格往往同时降低或升高。金融分析师则有兴趣知晓某只股票的价格变动在何种程度上受另一只股票价格变动的影响(总之,他们想知道股票价格之间的**相关性**)。如果这种联系足够强,则很可能这两个公司以某种方式相互关联着。贷款、持股、共享董事或股票价格的相关性是公司之间能够建立网络的主要标准:当出现这些情况之一时,则绘出网络图中的边。

互联互通远远超出了某个特定市场中的公司。正如金融危机所展现的,事件会迅速地从国家市场范围扩展为全球局面。这种情况能够发生的一个明显途径是国家之间的进出口贸易关系。这种**世界贸易网**是一种以国家为节点,以贸易关系为边的网络。就像细胞一样,经济依赖于这些网络的多层结构。

关键的基础设施

2003年9月28日晚,整个意大利的灯火都熄灭了,唯一的例外是撒丁岛。恢复正常供电耗费了数小时,有些地方甚至花费了数天时间。调查发现,这场停电由发生在意大利与瑞士之间高压线路邻近处的树木闪燃引发。由此造成的供电不足导致人们对剩余线路的供电需求猛增。结果，这些线路崩溃了，并在整个电力系统中产生了某种涟漪效应。

　　大规模的停电揭示了电网的连通性。这些系统跨越很远的距离将电力从中心点传输至城市和工业区域。一开始的电网规划很周密，这些电网随着时间的推移而越发复杂。如今，由高压输电线路连接的发电机、变压器和变电站组成的电力网络跨越数个地区，通常是数个国家（正如2003年的例子所展现的那般）。很明显，这种网络需要仔细维护以防止危急情况。

　　类似的不稳定性出现在各种其他基础设施中。电话网络这样的通信系统便是一例。但可能最敏感的是交通网络：街道、高速公路和连接城市的铁路，运输燃料和其他货物的船只网络，最重要的则是机场网络。飞机每年运输数十亿乘客以及成吨的货物。这种基础设施中的一次微小故障都会产生重大后果：据欧洲空中航行安全组织估计，航班延误给欧洲国家造成的损失仅在1999年就高达2 000亿欧元。在全球化的世界中，交通网络的作用类似于生物体内的循环系统。

如世界般巨大之网

　　1969年10月，一条信息首次通过电话线从一台计算机传至另外一台。加利福尼亚州的两个大学实验室位于这条电话线的两端。几个字母的传输之后，信息中断了，但连接业已建立：互联网的前身阿帕网（Arpanet）诞生了。关于计算机网络的设想在过去十年里一直都有。50年代末，美国国防部高级研究计划署（US Advanced Research Project Agency, ARPA）要求工程师保罗·巴兰设计能够抵抗攻击的通信架构。尤其是即便在其部分正遭受破坏的情况下,整个系统还必须继续工作。巴兰恰当地设计了具备这种特征的分布式系统——但策略上的调整使得他的开创性研究被束之高阁。然而,在60年代,一些大学为了其他目的而要求 ARPA 资助一个类似的项目。学术机构渴望相互连接它们的计算机,以便汇聚其计算能力。

1969年,阿帕网连接了加州大学洛杉矶分校和斯坦福研究所。两年之后,阿帕网的节点数已超过40个,包括一些公司以及其他大学。这种通信架构是如此成功,以至于在70年代,由粒子物理学家、天文学家和一些企业创建的类似网络,如高能物理网、Span网、远程网等纷纷出现在了世界其他地方。如果说一开始的问题在于连接计算机,后来则转向了如何连接各种网络。**网际互连**成为许多计算机科学家的座右铭。70年代末,工程师罗伯特·卡恩和数学家文顿·瑟夫开发了传输控制协议/互联网协议(TCP/IP):无论何种网络内部架构,这个软件都允许它们彼此交换信息。这个代码基于**开放体系结构**的概念,并被应用于公共领域。最终在80年代,**TCP/IP 转换**完全实现,互联网——"众网之网"——由此诞生。

这一结构很可能是最能体现网络思想的人工制品。一台连接到互联网的计算机即成为众多**主机**之一。如果要向特定地点发送电子邮件,我们并不需要直接与该目的地相互关联。从原点到我们的目标,信息沿着**路由器**这种负责传送数据包的设备传播。大量的连接使结构内部相互关联:光纤、电话线、卫星连接等。因为没人筹划过哪里应该增加主机和连接,人们并未记录互联网的大致结构。实际上,绘制主机层面的图景是不可能的，我们只能在路由器层面对其进行大致的呈现。在这种情况下，这些网络的节点便是路由器，它们的连接就是边。我们甚至可将结构更加粗粒度化，将路由器编组至**自治系统**中。这些组便是自主管理的域，通常对应于互联网服务提供商和其他组织。

互联网的巨大成功在于它提供的卓越体验。观看电视乃单向、单媒体的被动体验。互联网并非如此。人们可以浏览无数的文档，使用不同的媒体，交换信息，并相互交谈。与传统的通信技术，比如电话、无线电或电视不同，互联网并没有特定的目的。相反，它是一个能够容纳无数设备的异形人工制品。实际上，互联网仅仅是一个支持各种服务的物理基础设施。其中最成功的乃是万维网（WWW）。这是一组巨大的电子**文档**，被记录在组成互联网的设备之中，并通过那些为它们提供导航的**超链接**相互关联。这种模式有些类似于由文章、书籍、专利等组成，且经引用而相互关联的科学文献整体。

万维网的想法诞生于欧洲核子研究组织。物理学家蒂姆·伯纳斯–李（计算机科学家罗伯特·卡约后来也加入进来）于1989年提出了这一想法。伯纳斯–李设计了一个允许科学家们通过自己的计算机访问粒子物理实验所得出的大量数据的系统。让这一系统得以运转的软件并未申请专利，而是被公开发行。这一决定——就像TCP/IP的例子一样——被证明意义重大。从一开始，成千上万（之后甚至更多）的用户便对其进行了试用和改进，进而创建了各种网页和服务。短短几年之内，网络覆盖了整个世界。它的量级无人知晓，因为任何探索网络的搜索引擎（比如谷歌或雅虎）都无法归档所有网页。毕竟，这也没什么意义：几个网站一经请求就能生成新的页面。一项2005年的评估认为，整个万维网（对于静态页面而言）的内容相当于200太字节的信息。当时，这个数据量大约相当于10个美国国会图书馆。毫无疑问，如今这一数字会比之前多出若干数量级，因为万维网的信息量增长呈指数级。

赛博空间

2001年9月11日，纽约市的基础设施经历了一次"网络灾难"，这一灾难与当天发生的人类悲剧有关且类似。在两架被劫持的飞机撞向世贸中心大厦之后不久，记录显示通话数量激增。人们试图与朋友取得联系，并与同事一道救援相关人员。手机网络很快超载，人们纷纷在曼哈顿的付费电话旁排队等候。这次袭击破坏了威瑞森[①]的总部，切断了20万条线路。美国电话电报公司的基础设施，包括那些设置在世贸中心地下室的也遭到损毁。当电话呼叫失败时，许多人便求助于互联网。但无线服务也受到影响。而袭击对经济的影响则远远超出事故地之外。纽约证券交易所耗时六天才重新恢复运转，而各种服务直到数月之后才恢复至灾难前的水平。

这场恐怖袭击表明，几乎没有任何网络能够独立存在。物理和虚拟的基础设施已嵌入公共**赛博空间**（cyberspace），并在其中提供能量、信息、传输和通信等服务。电网支撑着互联网，互联网存储万维网，万维网又让邮件服务、社交网络、信息网站和文件共享系统运转起来。许多活动，包括飞行管控、银行项

---

① 美国无线运营商。——编注

目、应急系统和商业服务等也依赖万维网。赛博空间中某个层面的崩溃通常会以相当不可预知的方式影响到其他层面。

多个网络的相互连接在许多其他情况下也很常见。例如，影响2008年经济的流动性冲击迅速蔓延至其他许多经济网络之中。类似地，社交网络也在许多方面显示出这种特征。一个有趣的例子是真实世界的友谊与虚拟社交网络的联系相比较而显示的特征：这两个网络之间有着重要的反馈机制。细胞则是微型赛博空间：基因调控网络、蛋白质相互作用网络和代谢网络之间彼此嵌套。因此，一些科学家提出将基因组、蛋白质组、代谢组等相关概念融合为**相互作用组**这一综合概念。最后，生态系统可被看作是相互作用的网络组。例如，对抗和共生网络都在决定物种如何繁盛方面发挥着自己的作用。在所有这些情况下，网络为解开相互交织的复杂系统提供了有用的图景。

第四章
连接与闭合

同一个世界

2006年11月4日,德国西北部某地的一根输电线的断开引发了一场影响远及葡萄牙的雪崩式停电。人们以为这种小型断线故障的影响只限于区域层面,或至多限于国家电网范围之内。但电网的集成度已越来越高(目前,它们已在洲际层面形成了一个大型系统),因而也更为脆弱。其他基础设施的情况也一样。例如,几乎所有的机场都是相互关联的:如今,人们从任何起点出发都可以沿着经停点有限的路线抵达几乎任何目的地。互联网也是完全互联的,因为它本就是从更小的网络集成中成长起来的。

自然和社会中的网络可能在局部地区显得极为分散。细胞中的一些化学物质组仅仅在彼此之间相互作用,而不与其他任何物质发生作用。在生态系统中,某些物种群体会建立小型食物网,而与外部物种没有任何联系。而在社会系统中,某些人类群体可能也与其他人完全隔绝。然而,这种分离的群体,或者称**分量**,仅为极少数。在所有网络中,系统内几乎所有组成部分都会参与到一个大型的连接结构之中,这个结构被称为**巨型连通分量**。

例如,科学家在基于词汇联想的一个实验中,发现96%的语词止步于某个语词大群体内部。人们可以在这个词群中找到任意两个词语的关联路径,甚至像"火山"与"胃"这样差别巨大的语词之间也有关联:该实验的参与者所绘制的联想链为,"火山"—"夏威夷"—"放松"—"舒适"—"疼痛"—"胃"。一般来说,巨型连通分量在几乎所有网络中都会占到90%—95%的系统比例。我们可以列举这一事实的一些有趣后果。在性关系网络中,过去和当下的性行为能将我们与那些从未想象过或渴望有关系的人联系起来。大型的合作结构会出现在科学家的合作网络中,略去的只是少量孤僻的参与者。在公司的董事会中,连锁董事给予了绝大多数公司某种关联性。最后,在食物网中,某个给定地点的某物种所引入的污染物,会通过食物链而被带到远至地球另一端明显不相关的物种身上。

生活在一个大的相互连接的世界中并不意味着其中的任意两个节点都能够互相抵达。正如对普通道路而言,是否为单行道非常关键,我们也必须知道边是否有向。在有向网络中,从一个节点到另外一个节点存在路径并不保证这条路反向也能行得通。狼吃羊,羊吃草,但草并不吃羊,羊也不吃狼。这种限制在巨型连通分量中创建了某种复杂的架构(图6)。例如,根据1999年的估计,若忽略边的方向,万维网有超过90%的部分由相互连接的页面组成。然而,如果我们将边的方向纳入考虑范围,

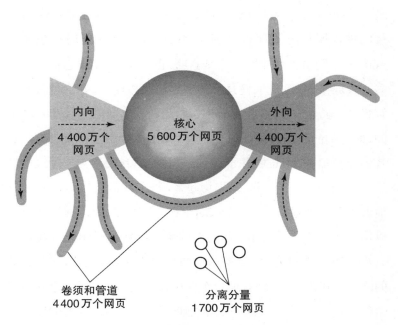

图6 像万维网这样的有向网络会显示出"蝴蝶结"结构:它由中间的巨型强连通分量、内向分量、外向分量以及各种小结构(管道、卷须以及少量分离分量)组成。图中为1999年的数据

万维网中彼此能够相互抵达的节点比例仅为24%,这部分构成了所谓的**巨型强连通分量**。其余部分则以**内向分量**和**外向分量**进行划分:前者由那些路径指向巨型强连通分量的页面组成,后者则由那些接收来自巨型强连通分量的连接的节点组成(图形由被称为管道和卷须的小型结构补充完成)。这种特征结构赋予了万维网这个巨型连通分量结构某种"蝴蝶结"的外形特征。这种复杂的结构并非局限于万维网,而是在所有有向网络中都会带着不同的分量出现。

巨型连通分量的存在——无论是否具备"蝴蝶结"结构——

是一个非常显著的特征。例如,尽管万维网的巨型强连通分量的规模比系统本身的1/3还要小,但它在1999年时仍包含了5 600万个网页。如果网络非常**紧密**,也就是说,如果它们拥有许多冗余连接,多到能够使几乎各节点都互相连接,那么便很容易解释这一特征。但通常情况并非如此。大多数网络反而是**稀疏**的,即它们的连接数往往很少。以机场网络为例:每个频繁旅行者的经验都表明,直航并不常有,抵达一些目的地需要经过中转站;活跃的机场数上千,但平均起来,每个城市仅与不到20个其他城市相互连通。大多数网络中的情况也是如此。这一情况可以由节点的平均连接数,即它们的**平均度数**体现出来。我们每天都创建许多网页,但每个网页的平均连接数大约为10。无数的路由器接入了互联网,但每个路由器平均只与不到3个其他路由器相连。最后,在超过5万名物理学家的样本中,人们发现合作者的平均数量仅大约为9。在大多数真实世界网络中,可能存在具有多连接的元素(事实上也确实存在),但一般说来,与此相对应的图并不紧密;相反,它们被认为是**稀疏**的。

  这个令人困惑的矛盾——一个稀疏的网络仍旧可以很好地连接起来——已经吸引了我们在第二章提到的匈牙利数学家保罗·埃尔德什和阿尔弗雷德·雷尼的注意。他们通过生成这些图的不同随机图解开了这个困惑。在每个随机图中,他们都改变了边的密度。他们以非常低的密度为起点:平均每个节点少于一条边。人们自然会期待,随着边的密度的增加,越来越多的节点将会彼此连接。但埃尔德什和雷尼所发现的却是一个相当突然的转变:一些分离分量突然合并为一个大的分量,后者几乎包含了所有节点。突变发生在边达到一个特定的临界密度值时：当每个节点连线的平均数（即平均度数）大于一时，巨型连通分量便会突然出现。这个结果意味着网络展现出某种十分特别的节约特征（这种特征是其无序结构所固有的）：甚至在节点之间随机分布的少量的边，都足以产生能够吸收几乎所有元素的大型结构。

近在咫尺

　　1994年初，奥尔布赖特学院的三名学生——克雷格·法斯、布莱恩·特特尔、迈克·吉内利——在一场暴风雪天气里看电视。据他们说，他们注意到电视上预告了凯文·贝肯的下一部电影，但他也同时出现在了许多不同的电影中。于是他们开始历数在那些电影中与他合作的大量演员。贝肯有点类似于"娱乐世界的中心"的想法便开始传播，进而变得尽人皆知了，基于此甚至出现了一个网页，即"凯文·贝肯指南"：这个搜索引擎提供了贝肯和人们可能会搜索的任何其他演员之间的关系。值得注意的是，如果人们输入来自早先商业电影中某位西班牙演员的名字，比如帕科·马丁内斯·索里亚，该指南网页就会发现一个十分紧密的联系：马丁内斯与路易斯·因杜尼一起演过《在西班牙度假》(*Veraneo en España*)这部电影；后者和伊莱·瓦拉赫一起出演过电影《先开枪，再提问》(*Il Bianco, il Giallo e il Nero*)；而瓦拉赫又与凯文·贝肯一起出演过《神秘河》(*Mystic River*)。这种与贝肯相关的合作链几乎能在人们所能想到的任何演员身上发现。

　　这个令人惊讶的特征让人想起一场科学家参与的游戏。随机图专家保罗·埃尔德什乃是20世纪杰出的数学家。科学家们衡量了自己与埃尔德什的合作以作为自己荣誉的象征：那些与他共同撰写过一篇论文的人的埃尔德什数记为1；那些埃尔德什合著者的合著者的埃尔德什数记为2；而埃尔德什合著者的合著者的合著者的埃尔德什数记为3；以此类推。然而，只有最小的埃尔德什数字才是骄傲的真正来源：超过500位科学家乃埃尔德什的直接合著者；而与这些核心的合著者合作过的科学家有几千人。最后，数以万计的人拥有埃尔德什数3（G.C.①就是其中之一；M.C.则拥有埃尔德什数4），所以这并非什么特别的荣誉。在任何领域几乎都没有科学家拥有超过13的埃尔德什数。

完全出乎人们的意料，这些引人注目的结果并非贝肯或埃尔德什特有。前者并非娱乐界的中心人物；后者也并非数学的枢纽。如果在任何其他演员或科学家身上重复同样的计算，也会发现类似的结果：非常短的链条关联着相隔甚远的个体。这一事实为一次十分普通的**鸡尾酒会经历**提供了某种有趣的洞见：当你与一位陌生人聊天的时候，突然发现他或她是你妻子的同学，或者你兄弟的网球搭档，抑或你朋友的邻居，等等。这种发现通常让人们惊呼（"世界真小！"），但可能它并非那般不寻常。社会系统似乎非常紧密地联系在一起：在一个足够大的陌生人群中，找到一对由十分短的关系链关联起来的人并非不可能。

六度分隔理论

1967年，美国心理学家斯坦利·米尔格拉姆进行了一系列值得铭记的实验。在杰弗里·特拉弗斯的协作下，米尔格拉姆

---

① G.C.及后面的M.C.即本书的作者圭多·卡尔达雷利与米凯莱·卡坦扎罗。——编注

向中西部（堪萨斯州和内布拉斯加州）的一些居民随机寄送了数十封信件。在信中，他请求这些人将信件转发至马萨诸塞州的某人（剑桥①的神学学生的妻子，或者波士顿的一位股票经纪人），而米尔格拉姆并未提供相关地址。如果这些居民并不认识收件人，米尔格拉姆则建议他们将信件转发至他们所认识的可能由于某种原因而与收件人"相熟"的人。人们在每次转发信件之时，都必须抄送一份给米尔格拉姆，以便他能够追踪信息的传递路径。在拥有数亿人口的国家内，依靠本质上为口口相传的方式找到某人似乎是不可能的。然而，几天后，收件人开始收到原始信件。这些信件仅仅经过了一位中间人。几周之后实验宣告完成，大约1/3的信件已经抵达其目的地：没有一封信件被转发超过10次，而平均邮寄次数为6次。

该实验激发了科学界的热情。1950年代，数学家曼弗雷德·科亨和政治学家伊锡尔·德·索拉·普尔便猜测人类之间的联系可能比预想的更加紧密。他们问道：如果从人群中随机抽取两人，他们认识彼此的概率有多大？通俗地讲，需要多长的相识链条才能将他们相互关联起来？在一篇最终于1978年发表的广为流传的论文中，他们提出了一个数学模型，表明在诸如美国这样的人口规模中，让人意想不到的是有很大一部分人两两之间都可以通过少量的中间人链条而彼此连接。米尔格拉姆的实验则是对此二人直觉的检验。这个发现有很强的甚至超出学术界的影响。**六度分隔理论**这个术语中的数字来源于米尔格拉姆的实验发现，并成为其结果的一个通俗的表达。1990年，剧作

---

① 此处指位于马萨诸塞州剑桥市的哈佛大学神学院。——编注

家约翰·瓜雷将其作为某喜剧的标题，该剧中一个魅力人物躲避人群，并声称自己是演员西德尼·波蒂埃的儿子。以下便是瓜雷表达米尔格拉姆实验结果的方式：

> 我曾在某处读到过。这个星球上的每个人和其他人之间都仅隔着六个人。这就是六度分隔理论。在我们和这颗星球上的其他每个人之间。美国总统。威尼斯的贡多拉船夫。还可以继续写下名字。[……]不仅仅是名人。而是任何人。雨林中的当地人。火地岛人。爱斯基摩人。我通过六个人与这颗星球上的每个人紧密地联系在一起。

小世界

"爱虫"（*Iloveyou*）病毒是史上最具传染性的计算机病毒之一。自2000年5月出现之后，它便感染了世界各地数千万的计算机，造成主要用于清除它的损失达数十亿欧元。"爱虫"病毒以电子邮件的形式传播，它伪装在一个酷似情书的附件中。当附件被打开，病毒便会感染计算机，并将自己转发至该计算机的地址簿中所含的电邮地址中。经过仅仅几次这样的复制，该病毒便感染了大量设备。正如前面某段落所描绘的社交网络一样，人们或许这样谈论病毒传播的计算机网络："世界真小！"仅仅数次连接便可访问大量计算机。彼此明显相隔甚远的计算机结果仅通过较短的关系链便相互连接。

在现实中，作为米尔格拉姆实验主要结果的这种**小世界属性**，存在于所有网络中。互联网由数十万个路由器组成，但信息包仅通过大约10次"跳跃"便足以从其中一个路由器抵达其他任何一个。两者之间可能相隔数千公里，但距离并不重要：重要的是必经的连接数，而这个数字往往很小。这就是为何信息能以非凡的速度在地球上传播的原因。另外一个例子是万维网：它由数十亿页面构成，但科学家发现，通常20次鼠标点击便足以连接其中任何两个页面。一般而言，在一只秀丽隐杆线虫的大脑中，任何一对神经元之间的距离都少于"三度分隔"。而在连接世界上所有国家的进出口网络中，无法找到被超过两条联系链条分隔的国家。这样的例子在其他很多事例中数不胜数。

小世界属性包含任意两节点之间的平均距离（被测定为关联它们的最短路径）十分小这一事实。给定某网络中的一个节点（比如在合著者网络中的埃尔德什），少数节点与它十分接近（即直接的合著者），与其相隔甚远的仍是少数（即那些持有很大埃尔德什数的科学家）；多数人则位于平均——以及很短——距离之内。这一点适用于所有网络：从一个特定节点开始，几乎所有的节点都仅与其相隔寥寥几步；特定范围内的节点数则随着距离的增加以指数方式快速增加。相同现象的另一种解释方式（科学家们通常的说明方式）如下：即便我们向某个网络中增加了许多节点，平均距离也不会增加太多；人们不得不将网络大小增加几个数量级后，才能注意到通往新节点的路径变长了（一点点）。

小世界属性对许多网络现象至关重要。新皮质中较短的突触距离对其功能可能非常关键：一些研究表明，神经退行性疾病如阿尔茨海默症便说明大脑中的小世界属性大量丢失。而性关系网络中的短程关系则表明，解释性传播疾病中的风险群体这一概念时必须小心翼翼：事实上每个人距离那些被感染者都很近。网络这种有效传播病毒的能力也有其建设性用途。采用**病毒式营销**策略的首批案例之一是微软1996年推出的电子邮箱服务（Hotmail）的全球性扩散。购买免费微软电邮地址的人须同意在他们的信件中附加一个链接，该链接允许收件人相应地打开一个免费地址。微软电邮在通信公司中增长最为迅速，它俘获了数千万用户，部分原因在于这个策略巧妙地利用了电子邮件网络的小世界属性。

捷 径

小世界属性是网络所固有的。即便完全随机的埃尔德什-雷尼图也表现出这种性质。相比之下，普通的网格并不会展现这一特征。如果互联网是一个棋盘状晶格，则任意两个路由器之间的平均距离将达1 000步的级别，而网络本身也会慢很多：不会有快速的网页浏览，也没有即时的电子邮件。如果科学合作网络是一个网格，则埃尔德什仅会有中等数量的合著者；尽管后者的合作者数量会大些，但规模仍旧适中，以此类推：特定距离内的个体数量不会呈指数级增长，而是慢许多。如果神经网络是一个晶格，则增加神经元的数量（比如，由于大脑的自然发育）将显著增加新皮层内的平均传递距离：矛盾的是，大脑的发育会让人变得不那么聪明（一些年轻读者可能会同意这一点）。

那又是什么使得网络与网格有所不同？网络中表现出的小世界属性为何又在晶格中缺乏？1998年，物理学家邓肯·沃茨和数学家史蒂文·斯托加茨试图回答这些问题。他们以思考十分简单且规则的结构为出发点。这个结构为节点构成的圆形,其中每个节点都与其最近和次近的邻点相连(图7左)。这种结构可能代表了一组偏远的村庄,其中每个村庄都与其邻近村庄交换货物,偶尔也会与邻村的邻村交换。在这种规则的结构中,产品从生产者那里传递至遥远村庄的消费者处可能要很长一段距离。

接着,沃茨和斯托加茨引进一个类似规则使得两个遥远村庄之间开放路径。在实际操作中,他们切掉了原始结构的某条连线,并将其与其他节点重新随机连接。顷刻间,一个村庄的居民便能与之前遥远的地区而非其邻村交换货物了。然而,从整体上看,仅有几个村庄受到了这种变化的影响,圆圈中的一些区域彼此之间仍旧相隔甚远。人们可通过计算重连之后节点之间的平均距离看出这一点:引进新的捷径之后平均距离仅稍微变短。这时,两位科学家开放了更多的"路径"(图7右)。每次重新连接之后,他们都会计算节点之间的平均距离:他们发现,在重新增加仅仅几处连线之后,平均距离明显缩短。少

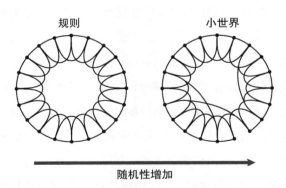

图7 在小世界网络模型中,引入无序后规则晶格便转换为网络,相应地,节点之间的距离也缩短了:小世界属性由此产生

量捷径便足以大大缩短系统中所有元素彼此之间的距离。将连接结构转变为小世界的关键因素便是少量无序的出现。没有哪个真实的网络能呈现出有序的元素排列。相反,其中总存在一些"不合群"的连接。正是由于这些无序连接的存在,网络方为小世界。

这些捷径在某些网络中很容易识别。例如,1858年第一个跨大西洋的电报电缆将欧洲和美国连接了起来。这个长达数千公里、重达数百吨的奇迹是儒勒·凡尔纳的《海底两万里》中潜水艇造访过的海洋奇迹之一。如今,数条跨洋电缆让信息能够在世界范围内即时传递。在语词网络中,一词多义乃捷径的主要来源之一。比如,"pupil"一词连接着两个语义区,一个是"教学"("pupil"作为"学生"),另一个是"视觉"("pupil"作为"瞳孔")。在社交网络中,格兰诺维特的**弱连带**概念——将不相关群体连接起来的联系——至少可以与沃茨和斯托加茨的捷径概念部分对应。

捷径在许多其他情况下也是引发小世界属性的原因之一。然而,我们不时也能找到另外可能的解释。例如,世界贸易网络中的超短距离则源于它是少数高度紧密(非稀疏)的网络之一这个事实:一个国家的贸易伙伴平均数量一般可与这个网络里的国家总数相当,这意味着每个国家都与其他大多数国家相互进行贸易。在食物网中,其他机制也能促成小世界属性。基位物种从阳光和周围环境处获得能量和物质,但它们的捕食者通常仅能摄取它们体内资源的10%,而在次第展开的每一步捕食中都是如此:如果食物链太长,顶端的捕食者便无法摄取足以维持其生存的资源。

无论最初怎样,当系统有了某种图形结构之时,小世界属性便成为需要考虑的重要特征。网络方法提供了这种系统的某种醒目景象:首先,系统的元素是某个大世界的一部分,其中几乎每个节点都具有彼此连接的路径;其次,这些路径非常短。这种交织的结构对理解从艾滋病到断电再到信息传播等广泛存在的现象而言至关重要。

第五章

超级连接器

枢　纽

在其令人印象深刻的"六度分隔理论"诸实验中,斯坦利·米尔格拉姆做了一项很久之后人们才能充分理解的观察。其中一个实验里,美国心理学家请一些随机挑选的内布拉斯加州公民转发一封信至马萨诸塞州的某股票经纪人。如果他们不认识收件人,就把信件寄送至那些他们认为与该经纪人更为接近的人。除了大部分信件平均仅用6步便抵达收件人处这一事实之外,米尔格拉姆还观察到,1/4的信件经由同一个源头传递至收件人,即该股票经纪人的服装商朋友,米尔格拉姆称之为雅各布先生。这一结果令人相当费解:为何如此之多指向该股票经纪人的关系路径都要经过这位先生?

经常坐飞机的人对类似现象都很熟悉。希思罗、法兰克福或纽约肯尼迪等机场对于环球旅行者而言都耳熟能详:无论目的地是何处,飞机都很可能会停靠这些机场。航空杂志常常附有世界地图，纵横交错的长线标出飞行路线：其中许多航线都会途经伦敦、法兰克福以及纽约等地，或将其作为终点。像这样的机场被称为**枢纽**，它们承载了整个航空交通的大部分运力。我们很容易得出，雅各布先生在社交网络中的位置与这些大型机场在空中交通中的地位相同。很可能，雅各布就是一个社交关系枢纽：他的许多联系人将其与其他一些人相互连接，所以很自然地，许多信件都要经他发出。

米尔格拉姆另一个引人注目的观察是，剩余大部分信件经由另外两人寄达：琼斯先生和布朗先生。借用空中交通这个比喻，这两人最可能是社交网络中"一般规模的机场"（如马德里或米兰等机场）。而剩余那些并非来自雅各布、琼斯或布朗的信件则途经社交网络中的较小"机场"（如吉罗纳或奥尔比亚等机场）。

这些枢纽的存在并不局限于该股票经纪人的社交网络或者机场网络。在许多其他系统以图表示后，其中也能看到类似的高度连接顶点或**超级连接器**。在许多网络中，人们都能看到"赢者通吃"的趋势：少数几个节点能吸引大多数连接，而余下的大部分节点将不得不共享剩余连接。现代分析表明，莫扎特创作的唐·乔万尼（此人引诱了2 065名女性，根据达蓬特的剧本："……在意大利640名，德国231名，法国100名，土耳其91名，而西班牙境内已达1 003名……"）这一人物并非夸张：在性行为网络中与他人连接最多的个人的性行为可达数千次。在一些数据集中，此类人中的一部分涉及性交易。自然地，这些与他人高度关联的个体最需要预防性传播疾病的感染。"9·11"恐怖袭击在纽约发生后不久，人们发现了超级连接器的又一例证：管理顾问瓦尔迪斯·克雷布斯画出了恐怖分子的社交网络简图,他发现这场阴谋的领导者之一穆罕默德·阿塔是连接最多的节点,即此人为该社交网络的枢纽。在科学合作网络中,我们也能发现那些与大量同行合作的关键人物:保罗·埃尔德什便是其中之一。

除了社交网络,超级连接器还存在于各种各样的网络中。互联网中的某些路由器具有数千条连接:这比一般的路由器多出几千条,后者仅有少数几条连接。**汇接机房**乃大型设施——通常为装满电缆的建筑物——数百个互联网服务商可经由它彼此连接:这些设施只要有一处出现故障就能导致整个区域(大到一个国家)失去网络连接。大型报纸的网站吸引了大量来自其他网站、博客和社交网络的链接。而在食物网中,位于顶端的物种捕食其他大量物种。最后,语词网络中的枢纽则为模棱两可或多义的语词:比如"arms",它在英语中既指身体四肢也指武器,因而与更大的语义场或同义词域相连。

枢纽也存在于细胞内的网络中。在基因调控网络里,单个基因可以控制大部分剩余基因组的表达:在某种细菌(新月柄杆菌)中,一个调节因子(CtrA基因)便可控制26%的细胞周期调节基因。p53分子是蛋白质相互作用网络的超级连接器:与该蛋白质相关的基因是强大的肿瘤抑制基因,它会在大量的肿瘤里发生突变。代谢网络的枢纽很明显是ATP分子(腺苷三磷酸):它在大量的生物化学反应中起到能量载体的作用。

巨人、侏儒和网络

从非利士营中出来一个讨战的名叫歌利亚的迦特人,身高六肘零一虎口。[……]头戴铜盔,身穿铠甲,甲重五千

舍客勒［……］铜戟枪杆粗如织布的机轴，铁枪头重六百舍客勒。(《撒母耳记上》，17：4—7)

根据《圣经》的《撒母耳记》，以色列人必须等待40天才有人敢于面对歌利亚那般强壮的人；随后大卫这个勇敢无畏的男孩进来了，他最终击败敌人。歌利亚并非寻常敌人："六肘零一虎口"的身高相当于大约3米高，而根据历史学家的推测，他身上"五千舍客勒"的铠甲也重达60—90公斤。

古代计量换算为现代计量时并不是很精确；而且，《圣经》的记述很可能是象征性的。然而，歌利亚的身高不是完全没有可能。根据《吉尼斯世界纪录》，有记录的最高者为一个名叫罗伯特·瓦德洛的美国人，此人身高2.75米。与歌利亚有着特制的盔甲和适合其身高的矛不同，特别高的人周围的物体与之相比都太过短小：椅子不舒服，天花板太低，他们还需要穿特制的鞋子和衣服。

他们的问题根源在于，身体尺寸是一种**同质**量值。进入电影院的人有着不同的身高，但所有的座椅都一样：有些人觉得椅子大了，其他人觉得小了，但一般而言，他们都觉得还算舒服。身体尺寸不会偏离平均尺寸太多。很高（或很矮）的人非常罕见，越高（或越矮）的人则越少见。几乎每个人都认识1.9米高的人，但仅有少数人认识2米高的人，而几乎没有人认识身高2.5米的人。人们在其他一些特征上也具有同质量值。例如，人们的智商测试结果多数时候接近平均水平，而偏差——无论往上还是向下——则较为罕见。人们的行为方式也十分同质化。比如，司机可能多少都有些莽撞，但经测量，多数时候他们在高速公路上的行驶速度都非常接近平均水平。

然而，同质性并非金科玉律。例如，一个人的朋友数量是极度多变的。根据《吉尼斯世界纪录》，瓦德洛的身高仅是最矮的人的五倍，后者名为钱德拉·巴哈杜尔·唐吉，身高55厘米。相比之下，最友好的人（即社交网络中的枢纽）所交的朋友数比那些仅与很少人交往的极度害羞之人要多出数十上百个。如果将虚拟社交网络中的联系人也算入一个人的朋友之中，那么这些网络的枢纽人物会比那些不善交往的人多出数百个朋友。人们的身高属于同质量值，但社交关系的数量却是**异质**的。

如果人的身高反映了他们社交关系的数量，那么，像瓦德洛这么高的人不会进入任何世界纪录。社交关系中会有比矮子高出几百倍的人：身高超过两公里的"社交巨人"会行走在社交之路上。更有趣的是，这些巨人在普遍矮小的人群中并不会是惊人的例外。侏儒和巨人之间的所有中间高度将由另一些人代表：自然，高度越高，人数越少；然而，这个想象世界里的高个子数量不会像在现实世界中那样迅速减少。换句话说，越高越少，但也不至于像在现实世界中那般稀有。

在这个想象世界中，座椅制造商的业务难度会增大许多，因为没法制造一个适合每个人身体尺寸的座位。而在现实世界里，如果想制造座椅、分析智商测试或预测自驾旅程的时长，我们会考虑平均身高、智商或行车速度。但为了理解社会关系，平均的概念就显得无用了。身体尺寸、智商、行车速度以及其他量值都具备**特征尺度**，即大多数情况下的平均值都是对我们所发现的实际值的大致预测。相比之下，社交关系并不具备这种尺度。如果去敲一个陌生邻居的门，你预计看见之人的身高会在一个合理的范围之内,而你的猜测多数时候是准确的。但我们几乎不可能提前猜测此人朋友数量的多寡以及具体数字。某个城镇的平均人际关系数量仅能让我们了解该地区社交网络的疏密程度。但我们无法据此对每个个人做出任何合理的预测。具备这种特征的系统被认为是**无标度**或**标度不变**的,意为该系统并不具备特征尺度。这句话还可以这样表述,相较于平均值,个体**波动**太大,以至于我们无法做出正确的预测。

肥尾效应

一般而言,具备异质连接性的网络都会有一组清晰的中心。当图很小时,我们很容易发现其内部连接是同质还是异质的(图8)。在第一种情况下,所有节点多少具备相同的连接性,而在后者中则很容易发现少量枢纽节点。但是,当被研究的网络非常大(如互联网、万维网、代谢网络以及许多其他网络)时,事情就没那么简单了。幸运的是,数学提供了一种方法来确定一种量值是同质还是异质。

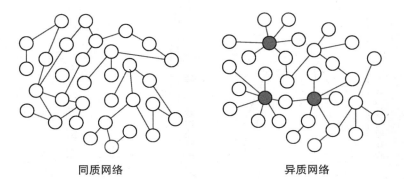

同质网络　　　　　　　　异质网络

图8　与存在高度连接节点(枢纽节点)的异质网络(右)相比,同质网络(左)中所有节点的度数大致相同

我们以同质量值为起点，比如人的身高。为了研究某班学生的身高，我们可以按照以下方法操作。首先，让那些身高在1.50到1.55米之间的学生排成一列：他们可能人数不多。然后，让那些身高在1.55到1.60米之间的学生平行地排成一列：这些人的数量会多些，队伍也会长些。接下来的一列为1.60到1.65米的学生：更多的人会出现在这一列。然后，每一列身高增加5厘米（图9左）。最后，这些列的轮廓将构成**钟形曲线**的形状：学生的数量随着身高的增加而增加，然后在平均值附近达到峰值，接着开始下降。很高和很矮的学生都较少，大部分处于中间范围。这条曲线提供了学生的身高分布。

现在，我们来考虑这些学生的社交关系数量。这时，每一列分别对应0到20个朋友，20到40个朋友，40到60个朋友，以此类推。该过程的结果提供了社交网络节点的连接性分布，即图的**度数分布**。这个结果图与身高图的情况十分不同（图9右）。首先，图中的列会更多，因为有的人的朋友数量会成百上千。多数人的联系人为几十个，但由此产生的分布将具有"肥尾效应"。

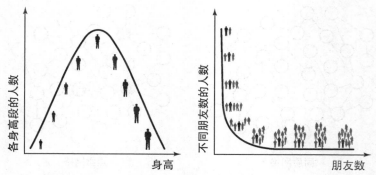

图9　人们的身高为同质量值，呈钟形曲线分布（左），而人们的朋友数量则为异质量值，呈幂律分布（右）

换言之，分布图的长尾或者说是"厚尾"将明显向右偏斜。从数学角度讲，度数分布的形状可通过**幂律**得到很好的描述。

在同质网络中，度数分布是类似于前述学生身高的钟形曲线，而在异质网络里，度数分布则遵循幂律，类似于朋友数量的分布图。幂律意味着异质网络中存在着比同质网络更多的枢纽节点（以及更多的连接数）。此外，枢纽节点并不是单独的例外：与连接数较少的网络相比，连接数较多的网络中有着完整的节点层级结构，每个节点都构成了一个枢纽。再拿身高和朋友数量来说。世界上身高1.50米的人可能有数百万；然而，如果我们将这一高度翻倍（即3米），如此高度的人则少得多，很可能没有。另一方面，数千万人在其社交网络中有比如说20位朋友。如果我们将这个数字加倍（40位朋友），拥有这个朋友数的人数则会少些（比如比加倍之前人数减少了1/4），但仍有数百万。我们可将这个数字多次加倍，而对应的人数则每次减少约1/4（实际减少的速度取决于幂律的斜率）。这便解释了比如琼斯先生和布朗先生在米尔格拉姆实验中的作用：虽然雅各布是股票经纪人社交网络中的最大枢纽，而琼斯和布朗是更小的枢纽，但与他们联系的人依然很多。

查看度数分布是检查网络是否为异质结构的最佳方法：如果度数分布呈肥尾，则该网络将有多个枢纽且为异质结构。人们从未发现某种数学上完美的幂律，因为这将意味着存在拥有无限连接数的枢纽。然而，不存在无限大的真实网络：这就是为何度数分布的肥尾总有一个度数最大值上限的原因。实际上，枢纽的大小会受到连接累积的各种成本的限制：例如，由于神经元的物理结构，神经元无法累积任意数量的连接。在专业协作网络中，时间起着某种作用：连接数无法无限累积，因为个体的事业（或生命）会在某个时刻终结。所有这些和其他因素都反映在度数分布的形状上。尽管如此，严重偏斜的肥尾状度数分布仍是异质网络的清楚信号，即便它从来都不是一个完美的幂律。

在解释枢纽和肥尾的含义时必须小心谨慎。例如，一些人类学家认为，一种叫**邓巴数字**的量值限制了人们的社交关系数量。根据这一假设，稳定的社交关系数量不能超过150这个数字太多。人类学家罗宾·邓巴在发现灵长类动物和人类的大脑皮质某部分的大小可能与它们的社会群组规模相关的证据之后，于1992年提出这一假设。如果这一假设为真，那又如何解释人们在许多社交网络中发现的有着上千联系人的社交枢纽呢？一些科学家认为，这便是"披萨送货员问题"的实例。披萨送货员在自己的手机上会接到许多电话，但只有极少的一部分来自其真正的朋友；其余则为客户。根据这种想法，呈现在社交网络度数分布图肥尾处的多数连接都是泛泛之交。然而，这还得看人们究竟想研究什么问题。例如，如果披萨送货员得了流感，流行病学家只会关心有多少人（不管是不是朋友）曾与他有过接触。

另一方面，并非所有网络都是异质结构。尽管小世界属性是网络结构所固有的，但不是所有的网络中都会出现枢纽。例如，电网通常就仅有少量枢纽。还有一些食物网、线虫的神经网络以及世界贸易网络等都很少有枢纽存在。

最后，人们在一些有向网络中发现了一个有趣的情况，正如在多数基因调控网络中发现的那样。如果基因A调控基因B，则箭头从A指向B，但B不是一定要指向A。**出度分布**（即朝外箭头数的分布)通常是肥尾状：少数基因调控大部分基因组。然而，**入度分布**(即朝内箭头数的分布)则均匀得多：少数其他基因调控某个基因。异质性在许多网络中都广泛存在，但当我们处理未知系统时，在检验之前我们不要理所当然地认为它就是异质网络。

自组织的标志

异质性和特征尺度的缺乏可能是无序的极好标志。推论如下。许多网络(比如互联网或社交网络)是在没有任何蓝图或监督的情况下成长起来的。因此，网络中的每个节点都遵循其自身的标准，并表现出彼此完全不同和不协调的行为。这些节点十分混乱，以至于它们很容易就被某种总体无序的过程所同化。因此，随机图应该成为这些网络的良好模型。这种推理似乎有效，但若深入检验，一些问题便出现了。最明显的是，随机网络根本不是异质结构。相反，它们的度数分布为钟形结构，这表明所有节点都有着大致相同的度数。随机关联节点的过程便是如此，即每个节点最终都有着同样的度数。更确切地说，度数具有特征尺度，且在平均值附近小幅波动。与许多真实的网络相比，随机网络中不存在枢纽节点。

其结果是，随机网络为小世界，而异质网络则是**超小世界**。也就是说，异质网络顶点间的距离小于其随机对照网络中的相应距离。如果人们在一个随机网络中加入一定数量的枢纽节点(因而使其更具异质结构)，则节点之间的距离会变小。相反，如果人们将一个异质网络随机化(即用相同数量的点和边建构一个网络，但让边随机分布)，枢纽节点便会消失，节点之间的平均距离也会变大。这表明枢纽节点承担了这些网络中的大多数连接：这些连接中的大部分正是从这些少数的超级连接点中产生的。

更重要的是，随机网络为同质结构这一事实意味着异质并不完全等同于无序。如前述随机网络那般的无序过程并不会产生那些在现实网络中所发现的异质连接。相反，异质性可能恰恰来自其对立面，即来自某种规则的有序行为。

这一点十分令人困惑，因为许多网络都不是设计的结果，也并非是在自上而下的严格监管中发展起来的。像电网或道路网络等少数网络由政治和技术权威控制，但大多数网络则无人监管。例如，互联网由地方一级的网络管理员控制，并且也受到技术、经济和地理特征的限制。然而，其大规模的结构在很大程度上则是未经筹划的：互联网非常类似于一项全球规模的实验，其中没人提出整体结构，而是靠无数代理人的行动将其建立。生物网络是一个更为清晰的例子：它没有设计者，仅有进化提供修补效果。社交网络、政治、金钱、宗教、语言和文化都会对个人之间的关系产生影响，但是当个人关系存在自由空间的时候，这些网络的形成便不再是严格规划的。在所有这些情况下，系统的整体组织产生于其组成部分的集体行为，即某种自下而上的**自组织**过程。这个过程可以解释为何许多网络即便没有规划，却仍然显示出异质性这样明显的有序标志。

异质性并非网络系统所独有。例如，地震强度便有肥尾状的分布特征，而如果人们绘制出地震频率与其强度的对比图，它便会呈现出很好的幂律分布。"平均强度的地震"并不存在，其中有很大的多变性，从无法察觉的震动到大规模的灾难。另一个例子是城市的规模：其范围从中国最大的大都会到托斯卡纳的小镇不等。还有就是收入分配：20世纪初，经济学家维尔弗雷多·帕累托指出意大利80%的土地掌握在20%的人手中。所有经济体中都存在不同程度的此种不均衡。

　　所有这些例子与网络共有一个基本特征：它们都是复杂且大体上无监管进程的结果。异质性并不等同于随机性。相反，异质性可能成为某种隐藏有序结构的标志，它并非某种自上而下的计划，而是由系统的所有元素共同作用而产生。这个特征在大范围不同网络中的出现表明，在许多这种网络中，可能有着某种共同的底层机制在起作用。了解这种自组织秩序的起源则是网络科学面临的核心挑战之一。

第六章

网络的涌现

永恒的变化

到20世纪90年代,互联网很大程度上还是一个未知的领域。尽管它已经成为通信、贸易和运输的关键基础设施,但没人清楚地知道其整体架构。管理员控制着本地网络,但他们并不清楚互联网的大规模结构。此外,互联网呈爆炸式增长:从70年代初的几十台机器到上千万台,且增长前景更为广阔。到90年代末,像康柏电脑公司(Compaq)和互联网数据分析合作协会这样的组织启动了一系列**绘图计划**,旨在探索互联网并描绘其全球布局。康柏公司的计划名为"墨卡托",以此纪念在16世纪绘制了最为重要的世界地图之一的地理学家,该地图将刚被发现的美洲大陆包含在内——而互联网被认为是一个尚待探索的"新世界"。多亏了这些以及其他项目,互联网地图才得以成为可能,它的增长现已受到监控。

但互联网的动态并未停止:此时此刻,路由器、计算机、电缆和卫星连接都仍在不断被添加或移除，这带来了持续增长的净效应。尽管缺乏规划，但互联网的增长过程也并非完全随机。相反，互联网是一种高度有序和高效的结构。这种涌现的秩序一定是构建网络之个体行为的某种规则性的后果。小规模的机制也必定存在，它们经由大量的交互迭代过程，最终产生了一种在大规模层面井然有序的结构。而解开隐含于网络结构形成过程之下的基本原理，对理解这些网络结构的自组织进程而言至关重要。

甚至那些看起来完全静止的网络，实际上也在进行着某种动态过程。细胞中的基因、蛋白质和代谢物所构成的网络，大脑中的神经网络以及生态系统中的物种网络等等，它们似乎都固定不变：基因调控、代谢途径、神经元连接或猎物–捕食者关系都相对稳定。然而，细胞内的网络在每个个体的发育期间呈爆炸式增长，并且随着生物体的老化及其对环境的应变过程而不断变化。大脑的可塑性在整个生命过程中可能会下降，但它并不会完全消失。物种灭绝或新物种入侵则会彻底重塑食物网。此外，所有生物网络在长期内都会因为自然选择的作用而发生改变。类似的事情还发生在其他明显的静态网络中，例如电网或语言网络。电网建成之后，由于事故和技术发展，它们会慢慢调整自身的形状。语词网络则随着说话者的改变而改变，并且，随着语言整体的演变，新词和新的语义关系也会被引入。

在其他网络中，变化主要集中在连接上，而顶点集合几乎不可改变。例如，某一国家的银行每天都会在银行同业网络中建立不同的借款模式。偶尔，该网络的顶点集合也能发生改变，比如在银行破产或者新的银行出现在市场上之时。然而，这种变化发生在比交易（即边的重新排列）更长的时间段内：在给定的一天中，网络中的变化主要与边相关。类似的事情不仅发生在股票间的价格相关性网络中（相关性的变化比实际的股票组频繁得多），也发生在世界贸易网络中（世界各国经济关系的变化比新的国家通过分离或联合而建立更为普遍），还有机场网络中（不管在哪一年，都仅有少数新机场开张，而大量航线则处于变化之中）。

完全相反的情况则出现在另外一组网络中：在这些图中，新节点被不断添加进来，而这一进程比连线的重新排列重要得多。这种情况最确切的实例是科学论文中的引用网络。引用较早论文的新论文每天都会出现，一旦发表，相应的引用关系便无法再次更改。

最后，在一些网络中，添加（或减少）节点和连线的动力会以相同的速度发生，并让位于一个相当复杂的过程。万维网以创建和删除新网页和新超级链接的方式不断更新。在维基百科这样的特定网站中，每天都会有新文章和文章间的新链接被创建出来。

可能的网络动力的范围十分广泛：网络科学家已经制定了各种衡量标准、数学模型和计算机模拟，以掌握这一过程背后的基本机制，以期理解这些网络的自组织原理。

富者更富

20世纪60年代初，社会学家哈丽雅特·朱克曼采访了一批获得诺贝尔奖的科学家，其目的是找到这些人工作方式中的独特之处，以及他们研究工作成功的秘诀。她在这些诺奖获得者的回答中发现了一个反复出现的主题。一位诺贝尔物理学奖获得者说:"在授予荣誉这件事上这个世界很奇怪。它往往将荣誉给予那些(已经)出名的人。"一位诺贝尔化学奖获得者补充道:"当人们在报纸上看见我的名字,他们往往会记住它,而忽略其他人名。"而一位生理学和医学奖获得者则具体谈道:"[当你在读一篇科学文献时]你常常注意到你熟悉的名字。即便排到最后,它也会让人印象深刻[……]你会记住它,而非长长的著者名单。""最出名的人会获得更多的声誉,在程度上毫无节制的声誉。"物理学奖获得者如此总结道。

这些观察促使另一位社会学家(朱克曼后来的丈夫)罗伯特·默顿于1968年提出了一个出色的定理。默顿提出,科学处于**马太效应**这种社会机制的影响之下。这个名字来自《马太福音》:

> 凡有的,还要加给他,叫他有余;没有的,连他所有的也要夺过来。(《马太福音》,25:29)

依照默顿的观点,这种机制在奖项、资助、关注度、声望等的分配中起着一定的作用。拥有大量此类资产的科学家可以轻松获得更多的类似资产。另一方面,那些缺乏者想要获取或保持这些资产就十分困难。

1976年,物理学家和科学史家德里克·德·索拉·普莱斯发现了支持这种观点的量化证据。普莱斯分析了大量因相互引用而彼此关联的科学文献。他发现,在特定时期内被大量引用的文献往往也会在之后获得更多引用,而那些一开始仅被少量引用的文献后来也未增加多少被引数。普莱斯从数学上证明了这个简单的法则能够解释高引文献(即引用网络的枢纽节点)出现的原因。更确切地说,他证明了这种机制能够解释为何每篇文章的引用数量分布显示出了幂律的肥尾特征。

普莱斯的模型是统计学家乔治·尤德里·尤尔和社会科学家赫伯特·A.西蒙早前开发的数学模型的变体,但是人们直到1999年才明白,这种机制可以解释网络枢纽、异质性、标度不变性在不同网络中整体以肥尾分布出现的原因。该模型的实现归功于物理学家奥尔贝特–拉斯洛·巴拉巴西和雷卡·奥尔贝特。巴拉巴西和奥尔贝特提出了网络增长的数学模型。他们设想了一个始于少量顶点(甚至仅有两三个)随机连接的图形。新的节点以稳定的速度加入到这个原初的核心中,每个节点都带有给定数量的连接。一个简单的规则确定了新节点是如何连接的:新节点会偏好已有许多连接的旧节点。这种机制又称**优先连接**(图10)。原则上,新节点可以加入任何旧节点之上,但旧节点的度数越高,它吸引新节点的概率便越高。偶尔,连接较少的节点也会接入新的连接,但大多数时候却是枢纽节点更具吸引力。

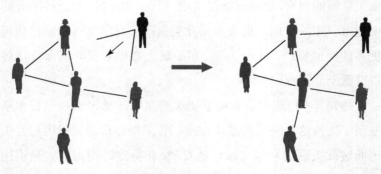

图10　在网络增长的优先连接机制中,新节点会优先连接度数高的旧节点

这一过程可从数学上加以研究,并能在计算机上进行模拟,人们可以测量真实的网络以确定其是否在起作用。一开始,所有节点多少都具备相同的度数。然而,随着节点的增长,有一些开始累积比其他节点更多的连接。某节点在特定时刻所具备的连接越多,它便越能够吸引新连接。这也是为何优先连接原则又被称为**富者更富**机制的原因。结果,连接性最初的小小差异会逐渐放大。因此,不同节点的层级结构便出现了,其中节点的度数差异极大,从连接最少的节点到那些累积了众多连接的枢纽不一而足。最后得到的网络具有异质性,且呈幂律度数分布。

巴拉巴西−奥尔贝特模型证明了,自下而上的增长机制能够产生异质性,而不用强加任何自上而下的设计。网络的全局标度不变性是个体及其局部选择的迭代结果:偏向连接更多的节点而不是更少的。这个模型利用概率以使个体在一定程度上可以偏离这种选择行为:一些节点可以决定连接度数低的节点。然而,总体的趋势决定结果。作为进一步的确认,我们可以检查在没有优先连接原则下成长起来的网络是否未出现异质性。的确,新节点会随机地连接到旧节点上,如此一来旧节点的度数并不影响自身吸引新连接的能力。结果出现了同质网络,其中每个节点都有着大致相同的度数。

普遍的机制

优先连接特指网络中的此种机制,而同样的机制还存在于许多自然和社会现象中,这些现象未来的演化取决于它们的历史。例如,城市规模依其目前的大小随时间而改变:大城市会经历较大的扩张,小城市则变化很小。明天的股票价格通常也与今天的价格构成一定的比例关系。这种机制也被称为**乘性噪声**。这一过程可能会在众多网络中起作用有着各种各样的原因。在某些情况下,具备许多连接是被新节点发现的主要方式。与许多其他网站连接的网站要比那些连接更少的网站更容易被浏览网页的人发现。高引文献中的情况也是如此。这种增强的受关注程度使得它更容易获得更多连接和引用。

在其他情况下,连接因其自身之故而具有吸引力。社会学家发现了**间接择偶**这一现象的证据,即人们选择自己的配偶不仅基于相关个体的个人特征,还会考虑他人意见。例如,一项研究发现,同一人在照片中被许多其他女性环绕时,要比其独照能获得处于大学年龄段的女性更高的评价。先前一系列合作伙伴的长名单可能会让一个人处于吸引更多合作伙伴的优势地位。这便是为何优先连接也被称为**受欢迎具有吸引力**原则的原因。

优先连接产生的一个更加有意的理由是,与枢纽节点相连能够使相关节点更容易接近许多其他节点。自1978年放松管制以来,许多航空公司纷纷采取了**枢纽政策**,此政策在于选择具备良好连接条件的机场作为它们最偏爱的目的地。动机很明确:由于这些机场与大量目的地相连,与之关联便会吸引更多的潜在客户。类似的事情也发生在连锁董事网络中:任职多个董事会的人可以获取大量信息,并具有广阔视野,这又使他具备很大的吸引力从而被雇用到更多的董事会中。访问在互联网中也至关重要。互联网的大部分由私人企业建立和维护,它们被称为**互联网服务提供商**。当它们中有一个建立了新的基础设施,其优先考虑之事便是允许用户快速访问网络中储存的信息。考虑到这一点,互联网服务提供商并不会随机选择它们想要连接的路由器，更不会只选择更近的。相反，它们会选择那些能够保证以最少的可能步骤连接到最多服务器的路由器。那么，又有什么比枢纽节点更能实现这一点？在互联网的情况中，绘图计划所提供的数据似乎符合优先连接的假设。数字显示，某一特定时刻发布的地图中带有很多连接的顶点，往往在下一幅图中会有更多连接。

在某些情况下，优先连接会以其他机制的面目出现。设想某人想要创建个人网页。常见的办法是查看朋友的页面，挑选一个不错的用作模板。由于人们常与其朋友有着共同的兴趣，因而新网页很可能会保留模板的大多数链接，或许会改变其中少数。所以，新的页面最终将是模板的副本，变化很少。现在，这种机制掩盖了优先连接的形式。事实上，模板网页会指向什么页面呢？它们很可能会指向枢纽页面。仅仅因为枢纽页面存在并俘获了万维网中的大部分连接，任一给定页面都更可能指向枢纽处，而非那些连接较少的页面。因此，如果某个页面被复制，则每个副本都会为枢纽带来更多的连接，从而形成有效的富者更富过程，优先连接规则因而得到恢复。复制机制看起来奇怪，但实际上它在某些情况下是主导因素。例如，一篇文章中的科学引文通常来自同一领域的其他文章，并往往会巩固该引用文章的权威性。

但复制机制最有趣的一种应用体现在基因网络中。基因组常常通过**复制和多样化**的过程演化。细胞复制时，所有的DNA都被复制到新细胞中，但有时候也会出错：原始DNA链条的一个完整基因被复制，并在子细胞的基因组中出现两次（**复制**）。大多数时候，这个新基因只是合成多余的蛋白质，这些多余蛋白质与另外那个相同基因所合成的蛋白质功能相同。然而,在进一步的复制中,两个相同基因中的一个可能会发生突变,从而会使自身合成的蛋白质行使新的生物功能,比如,与之前和自己一样的蛋白质之外的蛋白质发生反应(**多样化**)。许多情况中都已经发现了这种演化机制。现在,这种机制在蛋白质相互作用网络中的转化完全类似于复制机制:新的节点进入网络(复制和突变的蛋白质),它带来了一些前代的连接(即它最初与之反应的蛋白质),以及由突变而来的一些新蛋白质。高度连接的蛋白质在这种机制中占据天然优势:并非它们更可能被复制,而是它们比那些连接较弱的蛋白质更可能与复制蛋白质相关联,因而更有可能获得新连接。

尽管基因复制的作用仅在蛋白质相互作用的网络中显现,但代谢网络中也存在着支持优先连接的证据(直接或伪装)。这种关联过程的一个明显后果是,枢纽节点通常在网络里的旧节点中产生,因为它们之前有机会从"先发优势"中获益。现在,代谢枢纽正是可能在生命最初演化过程中并入基因组的原始分子,以及 RNA(核糖核酸)世界的残余物,比如辅酶 A、NAD(烟酰胺腺嘌呤二核苷酸)和 GTP(鸟苷三磷酸),或者最原始代谢途径的元素,比如糖酵解和三羧酸循环。在蛋白质相互作用网络的背景下,人们通过跨基因组的比较发现,平均来说,演化上较老的蛋白质与其他蛋白质之间的联系要比较新的蛋白质多。

优先连接并非网络中唯一起作用的机制,也并非所有的异质网络都起源于此。然而,巴拉巴西-奥尔贝特模型还是给出了重要的关键信息。某种简单的局部行为经过多次交互迭代作用,便能产生复杂的结构。这种结构是在没有任何整体设计下产生的,即便我们允许在行为中出现一定程度的随机性,它也会出现,这解释了就整体趋势而言的个体偏差。

适应度起作用之时

当偶然的性关系在紧要关头时,人们往往对潜在性伙伴的某些特征(例如政治观念、社会阶层或是否抽烟等)非常宽容。然而,当考虑订婚或结婚时,这些因素就变得十分重要。这是社会学家爱德华·O.劳曼对90年代中期的数据进行分析之后得到的信息。这些数据表明,美国约有3/4的已婚夫妇都具备大量的相似特征。其中包括,属于同样的社会阶层、种族群体,教育水平相同,甚至具有相近的魅力、政治观点和健康行为,比如饮食和吸烟习惯等。另一方面,当其他类型的性关系出现问题之时,相应的比例会低许多,尽管依旧很高(超过50%)。

同配生殖(即相似个体彼此结合的倾向)甚至在那些婚姻非包办、理论上任何人都能与任何其他人结合的社会中都是很强的趋势。在躁动的大学时代,个人的声望——例如,对此人之前的性伴侣数量进行衡量得出——可以成为塑造性关系网络的相关驱动力。但是,人们安定下来后,更严格的标准便会发生效力。尽管富者更富能在第一种情况中发挥作用,但它几乎无法解释第二种情况。同配生殖是**同质相吸**的一个具体例子:大量社会学证据表明,它是由相似个体相互关联所构成的一般趋势,也是塑造社交网络的强大力量。在巴拉巴西-奥尔贝特模型中,连接到一个节点的主要标准便是该节点的连接数,但在许多情况下,其他特征对吸引新连接而言重要得多,不管实际的连接数为多少。

巴拉巴西-奥尔贝特动力机制的一个结果是，旧节点相对于新节点而言具备累积优势。然而，实际情况并非总是如此。举个例子，万维网的旧日荣光，比如麦哲伦或Excite等搜索引擎大多已被遗忘。谷歌或雅虎等后来者已取代了前者的位置。累积优势在新的参与者进入游戏时可被完全推翻（想想脸书的情况）。新手往往具备一些固有特征，这让它们比之前的参与者更具吸引力。在这种情况下，网络的连接性并非完全由节点度数驱动，正如巴拉巴西-奥尔贝特模型中所展现的一样。相反，每个节点的某种特殊的特性能够在节点获得连接的能力中发挥十分重要的作用。这种特性被称为节点的**适应度**，或称为节点的**隐变量**，它能够塑造网络的结构，却不像连接数那般显而易见。

2002年，物理学家圭多·卡尔达雷利、安德里亚·卡波奇、保罗·德·罗斯·里奥斯以及米格尔·安琪儿·穆诺兹共同设计了一个仅凭节点适应度便可产生网络的模型。其基本思路与随机图相同：在一组节点内，考虑所有可能的配对，再依据某一给定概率看配对节点之间是否能够连接。然而，在这种情况下，概率并不固定，而是依节点在配对中的适应度而改变。模型首先会为节点分配适应度。比如，这个隐变量可能代表了个人的收入。在此情况下，适应度在节点间的分配便会类似于国家的财富分配：先是少数十分富有的节点，然后是一定数量的上层中产阶级节点，之后便是下层中产阶级节点，诸如此类。定义连接的概率是第二步。若要模拟高度隔离的社会，则可建立以下规则：两个人形成社会关系的概率取决于他们收入的差异；具体而言，这种概率与收入差距成反比。也就是说，收入差异越大，两人获得连接的概率便越小。两个收入差别很大的人之间也可能建立联系,但这绝不是社会的主流趋势:一般而言,同质相吸原则将最终胜出。

适应度模型这种机制可能看起来过于简化,但在某些情况下,它能完美地发挥作用:例如,在世界贸易网络中便是如此。在这种情况下,适应度便是世界各国的国内生产总值(GDP)。连接的规则如下:两个国家的适应度越高,它们彼此关联的概率就越大——这将是某种**适者更富**的机制,其中那些GDP极高的国家会在彼此之间建立更多的商业联系。这种机制与同质相吸原则并不相同,因为,尽管GDP高的国家之间会建立许多联系,但GDP较低的国家并不会与其他低收入国家建立联系:它们的联系仍旧少得可怜。物理学家迭戈·加拉斯凯利和马里耶拉·罗弗雷多于2004年证明,这些组成部分足以精确预测世界贸易网络的特征,只要人们在模型中写入任意时刻的已有商业联系总数。例如,该模型精确地预测到了网络的节点度数分布态势。这表明,隐含在世界贸易网络这种自组织之下的基本机制已被适应度模型这种简单的动力机制捕捉到了。

跟优先连接一样,适应度机制不大可能在所有的现实世界网络中起作用。巴拉巴西-奥尔贝特模型被用于扩张中的网络时还算可靠,适应度模型对静态网络也有效,后者的节点数大致是固定的。然而,这两种模型还能同时起作用:2001年,物理学家吉内斯特拉·比安科尼和巴拉巴西共同将适应度这一概念引入到优先连接模型中,证明了这两种效应的共同作用良好预测了网络的拓扑属性。但必须注意的是,适应度模型并不总呈幂律度数分布。大范围的适应度和连接规则才能产生幂律度数分布,但很多其他的则无法做到这一点。然而,这并非某种局限性,而是这种模型积极的一面,这使得它能应用于像世界贸易网络这样的非异质网络。

各种策略

邻家女孩(或男孩)的神话已成为历史。根据社会学家米歇尔·博宗和弗朗索瓦·埃朗1989年的一项研究,80年代中期,在自己小区找到配偶的人已微乎其微了,在法国这一比例仅为3%。然而,这种情况仅在30年前还十分普遍:博宗和埃朗发现,这一比例在1914年到1960年间为15%—20%。世界上许多国家中的情况依旧如此。在某些情况下,婚姻(以及一般意义上的社会关系)既不受某种流行趋势驱动,也不受相似性标准驱动:倘若地理限制影响很大(比如,无法定期搭乘远程交通),人们便会被迫与邻里以及同乡交友。

在这些情况下,网络的顶点被嵌入到物理空间中,这会带来许多重要后果。有时,人们几乎可以不费任何代价地与任何人建立联系(比如在虚拟网络上结交朋友)。但在其他情况下,远程联系则代价高昂。这两种情况下的网络属性也十分不同。许多基础设施网络(火车、煤气管、高速公路等)会显示出一种偏差,因为它们嵌入的是物理空间。其他网络则嵌入到时间之中:例如,科学论文会在某天发表,这会在网络连接中造成某种偏差,即新论文只能引用旧论文,而旧论文却无法引用新论文。

其他偏差和策略能够影响网络的形成。社会学家确定了人们在社交网络中建立联系的两种基本激励机制:一为**基于机遇的前因**,即两个人会建立联系的可能性,二为**基于利益的前因**,即某种促使关系形成的效用最大化或不适感最小化机制。某种数量上的全局最优化对于塑造技术网络能够发挥重要作用：例如，使万维网搜索成本最小化的压力会导致人们倾向于对最短路径长度和连接密度进行优化。

最后，甚至还可能出现这种情况：表面上的自组织起源于完全的随机状态。设想某公司发布了新的社交网络，并为10万人提供了昵称。然后，该公司许可1个人与其他1000个人建立联系，2个人可与其他500人建立联系，3个人可与333人建立联系，4个人可与其他250人建立联系，以此类推。人们并不认识昵称背后的人，所以，他们会随机地选择伙伴。显然，这一过程中并不存在自组织：该公司建立的规则决定了网络的结构。然而，最后形成的网络**在结构上**仍呈现出幂律度数分布。这个例子表明，在某些情况下，幂律并不总是意味着自组织过程。

若人们试图通过为其行为建模的方式来理解网络的特征，则策略、偏差、过程和动机等一系列广泛的因素都必须被考虑进去。甚至，每个个体网络都需要自己的模型。然而，一些十分普遍的机制，比如优先连接或与适应度相关的动力机制，都可能会在许多明显不相关的众多网络的形成过程中发挥作用。本章描述的模型简单地解释了，在缺乏全局规划的情况下，局部机制何以确实能产生规模庞大、复杂、有序及有效的结构。

第七章

深入挖掘网络

谁是你的朋友?

根据上世纪90年代开展的一些研究,每有一个患有性传播疾病的美国白人,在美国某些地区就有多达20名同病相怜的非裔美国人。持续的种族不平等导致了这一结果。然而,产生如此巨大差异的真正传染机制在一定程度上仍晦暗不明。1999年,社会学家爱德华·O.劳曼和尤思科·尤姆发现了一个有趣的证据:性活跃程度较低的非裔美国人(过去一年仅有1位性伴侣者)与性活跃程度更高的非裔美国人(过去一年中有4位或更多性伴侣的人)发生关系的可能性是相同情况下白人的5倍。换句话说,在白人的性关系网络中,不那么活跃的**外围群体**某种程度上与活跃的**核心群体**彼此隔离。相反,这两个群体在非裔美国人中的关联则更多。这种差异的原因尚不清楚,但其结果却很明确:在第一个网络中,性传播疾病主要在核心群体内部蔓延,而在非裔美国人中,这些疾病也溢出至外围人群。

在本例中，网络节点的度数对于理解这个现象并非最相关的变量。性伴侣数量相同的个体受感染的可能性也不尽相同，这取决于该个体是白人还是非裔美国人。在这样的情况下，仅仅知道你有多少"朋友"（即你所在节点的度数）还不够，还必须知道你的朋友有多少朋友。度数分布为图的大体结构提供了大量信息，比如它是否包含枢纽节点等。然而，度数分布并不能显示图的所有信息。比如，设想两个图具有相同数量的节点和边数：其中的节点可能有着完全相同的度数，但边的分布却可能导致这两幅图完全不同。度数乃顶点的局部特征。想要更加细致地认识网络结构，人们必须深入挖掘，并找到方法来描述节点的周边情况：距其最近的邻点，其邻点的邻点，等等。

在白人的性关系网络中，低度数节点往往与低度数节点相连，高度数节点则与高度数节点相连。这种现象又名**相称混合**：它是同质相吸的一种特殊形式，其中连接数类似的节点往往会互相连接。相反，在非裔美国人的性关系网络中，高度数节点和低度数节点则更容易彼此连接。这被称为**不相称混合**。这两种情况都显示了相邻节点在度数上的某种相关性。当相邻节点的度数呈正相关，则为相称混合；反之则为不相称混合。

通常，这些混合模式的存在是网络中某个重要机制作用的结果，这个重要机制可能就是自组织。在随机图中，给定节点的邻点完全是随机选择的：结果，相邻节点的度数之间并没有明确的相关性（尽管图的有限大小可在某种程度上掩饰这一点）。与此相反，大多数真实网络中都存在节点相关性。尽管不存在一般规则，但大多数自然和技术网络往往为不相称混合模式，而社交网络则为相称混合模式。例如,高度连接的网页、自主系统、物种或代谢物常常与其所在网络中连接较少的节点相互关联。另一方面,公司董事长、电影演员和科学文献作者往往与那些连接性与自己类似的人相关联:个体的节点度数越高,其网络邻居的度数也越高。

度数的相称和不相称仅为让节点相互关联发生偏差的大量可能相关性中的一个例子。例如,劳曼和尤姆也证明了,相比于其他群体,有更多的非裔美国人倾向于从自己的社区选择伴侣。因此,当感染进入社区之后,它便被"困"在里面了。单单这种简单的效应便让非裔美国人感染性病的可能性高出美国白人1.3倍。在这种情况下,相关性并非源于节点度数,而是与每个节点内在特性相关的一种同质相吸,这个内在特性即种族身份。另一个例子是体重的相关性:研究发现,相对于与自己体重指数不同的人而言,体重指数相似的人倾向于更频繁地在彼此之间建立社交联系。要注意的是,相关性并不必然就是支持同质相吸的正面因素:比如,在食物网中,边将植物与食草动物、食草动物与食肉动物相连,但绝少将食草动物与食草动物或者将植物与植物相连。

谁是你朋友的朋友?

科西莫·德·美第奇于15世纪带领家族接管佛罗伦萨,人们称他为"难以理解的斯芬克斯"。尽管他绝少公开发表言论,并且也从未公开采取任何形式的行动,但他依旧能够在自己周围建立起强大的党羽,并让自己成为文艺复兴时期最重要城市的国父(*pater patriae*)。1993年,社会学家约翰·F.帕吉特和克里斯托弗·K.安赛尔分析了美第奇家族与佛罗伦萨其他权势家族之间的婚姻关系、经济联系和赞助往来。他们发现,科西莫的家族位于众多权贵家族关系网络的中心。更重要者,若无美第奇家族搭线,其他家族多数时候的联系并不多,甚至彼此抵牾。科西莫的克制态度帮助自己建立起了与各家族的联盟和共治关系。

以美第奇家族为中心的网络便是**自我中心网络**的一个实例,在这种网络中,一组节点与中心节点(自我节点)直接连接,这组节点彼此之间也相互连接。每当后一种连接丢失一个(也就是自我节点的两个邻点彼此不再相邻),该网络就会出现**结构洞**。科西莫的网络布满了结构洞,其家族能够用它们实行分而治之(*divide et impera*)策略:美第奇家族被视为许多冲突的第三方,那些家族不得不要求美第奇家族调节他们彼此的关系。

然而,对个人而言,周围布满许多结构洞并不总是件好事。根据2004年的一项研究,朋友之间不构成朋友关系的青春期女孩,其自杀概率为相反情况的两倍。这个发现的可能解释是,当事人会暴露在无关朋友的冲突之中。另外一个例子来自工会:若工人之间的联系网不存在结构洞(即**自我节点**被大量相互关联的节点包围时),则会形成一个强大、协调良好、交往密切的组织。一般而言,结构洞的不同模式表示不同的情况。例如,专业领域的科学家常常与该领域其他科学家相互联系,后者可能彼此也有联系。另一方面,高度跨学科领域的科学家则很可能与不同领域的科学家都有联系,后者并不必然彼此关联。

在所有这些情况下,你有多少朋友(即你的节点度数),或者他们是谁(比如他们的节点度数与你的相似还是不同)都不重要。重要之事在于,你朋友的朋友是谁:特别是,你的朋友彼此之间是否也是朋友关系。这个概念通常被称为**传递性**或**集聚性**。让我们考虑有着两位朋友的一个人：他们三人构成了**连接三元组**。如果此人的两位朋友彼此也是朋友，那么他们三人同样构成了**可传递三元组**，或者**三角形**。网络中的三角形数量与其中的连接三元组总数的比值便是该网络**集聚系数**的基本组成：这个系数衡量了该图中的三角形密度及其总体的传递性。而在随机网络中，某节点最近邻点之间的连接与任意其他两个节点间的连接具有同样的随机性。因此，这些图仅有纯粹随机连接的边所组成的三角形。另一方面，几乎所有现实世界网络的集聚系数都高于其相应的随机网络。这意味着某种重要的过程——可能是某种形式的自组织——在产生这种额外的传递性过程中起到了作用。

许多网络的高度集聚性表明存在着其中"每人都是其他每个人的朋友"这样的群体。乍一看，这幅图景似乎与网络的小世界属性相矛盾：网络是否是个"开放"世界，其中每个人之间都仅隔几步之遥？或者说网络就是紧密编织的分离群体的总和？现实中，这两个特性之间并不存在真正的矛盾：通过仔细观察沃茨–斯托加茨模型就能看出这一点。这个模型以一圈节点为开端，每个节点都与其最近和次近的邻点相连，就像遥远的村庄与其邻村交换货物一样。这是个完全集聚的结构，其中，任一村庄的所有商业伙伴也互为商业伙伴。然后，该模型为随机选择的节点开放一些连接：少数村庄开放了前往其他遥远村庄的路径，并将货物带往该处，且拒绝与其邻居做生意。少量路径便足以陡然降低任意两个村庄之间的距离，但另一方面，我们可以认为当地紧密的商业结构已遭到破坏，也即，当地的集聚性下降了。然而，沃茨和斯托加茨发现,集聚性的降低并不像平均距离的减少那般明显。实际上,为了使传递性显著下降,人们必须重连几乎所有节点。此时,网络中仅剩随机连接了,而我们并不会期待随机图中存在高集聚性。此处的关键信息在于,网络(既不是有序网格,也非随机图)可以同时具备高集聚性和较短的平均节点距离这两个特征。

集聚性的另一个有趣之处在于,几乎所有网络中某节点的集聚性都取决于该节点的度数。通常,节点度数越大,集聚系数便越小。低度数节点往往属于彼此连接良好的局部网络。类似地,枢纽节点与众多节点连接,这些节点并不会直接地彼此连接。例如,互联网中低度数的自治网络常常属于高度集聚的区域网络,且经由全国主干网彼此连接。许多网络中都可能出现类似的结构,其中集聚性会随着节点度数的增加而呈下降趋势。

谁是你朋友的朋友的朋友……?

金钱会在一定程度上带来幸福,但周围快乐的人却能给人们带来更多的幸福。根据1984年的一项估计,每年多挣5 000美元会增加2%的幸福感,而社会学家尼古拉斯·克里斯塔基斯和詹姆斯·福勒2008年的一项研究表明,拥有一个快乐的朋友则会提升人们15%的幸福感。两位社会学家调查了来自马萨诸塞州弗雷明汉的超过12 000名居民的主观幸福感。不仅如此,他们还绘制了这些人之间的朋友、配偶或兄弟姐妹关系。通过绘制这个关系网,二人发现,联系紧密的人常常有着类似的感觉:幸福的人往往聚在一起,另一方面,不幸之人亦是如此。克里斯塔基斯和福勒甚至还发现了更多有趣的证据。个人的幸福会受到其非直接相邻的人的幸福程度的影响。两步之外(朋友的朋友)的"幸福效应"约为10%;三步之外(朋友的朋友的朋友)则为约6%。这种效应仅在第四步就消失了。这两位社会学家和其他科学家也在肥胖、吸烟习惯以及口头建议(比如找一个好的钢琴教师或一个好的宠物之家)等方面发现了类似的结果:在所有这些情况下,三度空间之外的影响或信息会对个人起作用。人们在不同社会过程中发现的这个**三度空间规则**是**超二元扩散**的一个例子,即超越连接最近邻居的**二元**关系的扩散现象。在这种情况下,每个节点的度数、其邻点的度数,还有邻点之间的连接都不再重要。影响超越了各节点最近的连接圈。实际上,在许多现象中,这种影响甚至超过了三度空间。例如,高度传染性疾病可形成更长的传染链;类似地,营养物则可扩散至整个食物网。

在这种动力机制中,节点的重要程度取决于通过它的链条数量。为了捕捉这个观点,社会学家林顿·C.弗里曼引入了节点的**中介中心性**这一概念。取某网络中的所有节点对,数一数关联它们的最短路径数。节点的中介中心性基本上就是穿过该节点之最短路径占所有路径数量的比例。这一比例越高,相关节点的中心性就越高。按照这种测量方式,美第奇家族便是15世纪佛罗伦萨家族中最具中心性的一个。在这种情况下,中介中心性便可衡量减缓节点流或扭曲通过链条的可能性,它以这种方式服务于中心节点的利益。一些研究表明,企业在经济网络里的中心性很好地预示了它的创新能力(根据获得的专利数衡量)以及财务业绩。有趣的是,1980年到2005年间,东亚国家在世界贸易网络里的中心性经历了大幅增长,而大多数拉美国家却呈下降趋势。然而,这两个地区的贸易统计显示出类似的模式：宏观经济统计并未很好地追踪二者发展的巨大差异，而基于网络的方法却捕捉到了这一点。根据1965年的一项研究，莫斯科在中世纪便成为俄罗斯中部河流运输网络最具中心性的节点。很可能，这为其未来的重要性打下了基础。

中心节点通常充当桥梁或瓶颈：它们几乎是网络交通中的必经站点。因此之故，中心性乃是对网络节点的负荷的估计，前提是大多数节点之间的连接都经过最短路径（情况并非总是如此，但这是个很好的近似情况）。由于同样的原因，中心节点的损坏（例如，某个中心物种灭绝或某个中央路由器被破坏等）会从根本上影响相关网络的节点连接数。根据想要研究的过程，还可以引入中心性的其他定义。例如，**接近中心性**计算某节点到其他所有节点之间的距离，而**抵达中心性**则将网络内所有节点分解为经由一步、两步、三步等不同步数抵达的节点。此外中心性还有一些更为复杂的定义。

许多现实世界网络的中心性特征是它们异质性的深层标志。许多真实世界的网络会表现出异质分布特有的长尾。平均中心性并非对任一节点的有效估计，因为这一量值在平均范围内变化很大：少数节点便是网络中几乎所有最短路径的主要瓶颈，整个中心性较低的节点层级都要经由他们向下。考虑到中心性较高节点的重要性，我们很自然地会问它们是否与网络的枢纽节点相同。很多情况下，事实的确如此。例如，高度连接的自治系统也可作为区域网络的桥梁；多义词因与许多其他语词连接，从而将语言中的不同领域联系在一起。但这并非普遍规律。机场便是一个显著的例外：其中，某些低度数机场具有特别大的中介性。2000年，中心性最高的机场为巴黎机场，它是关联着250多座城市的枢纽节点。中心性次高的机场则为位于阿拉斯加州的偏远的安克雷奇,它是一个仅与40座城市连接的中等大小的机场。其他类似机场也出现在最具中心性的机场名单上。这种异常又该如何解释?阿拉斯加州有许多飞国内航线的机场,但安克雷奇是通往美国其他地方的唯一桥梁,因此,许多航线都穿过该机场。这种异常是局部地区机场密度高,却少有通往国外的航线的结果。

你属于什么群体?

1972年,美国一所大学俱乐部的两名空手道教练发生了激烈的冲突,从而决定将他们的俱乐部一分为二。这件在世界上多数人看来不值一提的小事却成了社会学家韦恩·W.扎卡里眼中的一座金矿。1977年,他发布了一项关于此事的开创性研究,并在其中提出了一个新奇的观点。

1970年,空手道教练希先生要求俱乐部主席约翰·A.提高课程价格,以提供更好的薪酬。而他所得到的只是拒绝。随着时间的推移,整个俱乐部都因此事而产生了分歧,两年后,希先生的支持者们在其领导之下组建了一个新的空手道组织。在此期间,扎卡里搜集了空手道课程、会议、派对以及俱乐部成员的聚会等相关信息,并将那些在俱乐部之外还见面的人定义为好友。这时,他便能为该俱乐部绘制一幅精确的友谊网络图谱了:所得图形的结构明显围绕两位教练而分为两个群体,每个群体里的人都互为好友且与其中一位教练交好,而两个群体的成员之间却很少往来(图11)。当俱乐部一分为二,人们几乎都沿区分两个群体的界线站队。

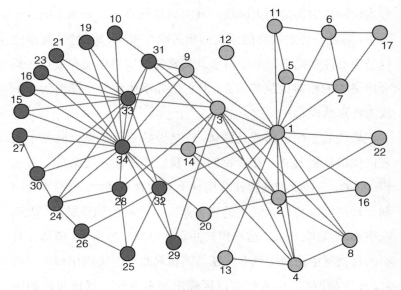

图11 人类学家韦恩·W.扎卡里研究的空手道俱乐部中的友谊结构能让我们预测到该群体会一分为二

仅仅基于网络结构本身,扎卡里的方法便能够几乎完美地预测俱乐部的分裂。从那时起,研究人员便一直致力于找出能够识别网络中的**社区**或**模块**的通用办法。在扎卡里的例子中,这些社区或模块仅通过查看图形便能看出,但在其他复杂得多的情况中,人们尚未发现通用的解决方案。所有真实世界的网络都在一定程度上显示出模块化特征。很明显,阿拉斯加机场便是机场网络结构中一个特定的模块,其他内部连接良好却与外部没有连接的区域也是如此。食物网也分为若干不同的**分部**,即那些内部互动更加频繁而与其他物种联系较少的物种群体。社交网络也分为不同的**团体**:例如,针对青少年的研究表明,他们的行为强烈地受其所属群体的影响。神经网络常被划分为对应特定功能的大区块。基因调控网络则被分为不同的子网络，后者与特定的功能或疾病相关联。度数、相关性、集聚性以及中心性都提供了单个节点及其紧邻的周遭环境和节点在整个网络中的相对地位等信息，但它们并不体现整个图形所分解成的各别结构。

模块的更简单形式是**模体**，它是少数节点在整个网络中重复出现的连接模式。在食物网中，我们经常会发现某种菱形结构：例如，某种食肉动物捕食两种不同的食草动物，后两者食用同一种植物。另一种常见的模体是三个物种的简单链条：大鱼吃小鱼，小鱼吃虾米。这些模式并非纯粹概率的结果：模体在真实的食物网中出现的频率比在其随机对照网络中高出很多。通常，在大型网络中，人们可以区隔出许多可能为候选模体的由节点和边组成的子集。然而，只有当给定的子图出现在某网络中的频率高于其随机对照图时，该子图才能被认为是相关模体。在万维网中，一个十分常见的例子是**二分团**：它由两组网站组成，其中一组的所有网站与另一组所有网站之间相互连接。通常，这种模体能够确定一组有着相同兴趣的"粉丝"群体（比如有关漂流的博客），并指向他们的"偶像"（例如漂流杂志的网站等）。调控基因的网络则几乎完全由模体构建。当大肠杆菌处于应激状态时，特定的基因回路会感知到应激状态，并协调某些蛋白质的产生。这些蛋白质联合起来形成鞭毛，这种不断摆动的尾状物能让细菌游走以寻找更好的环境。相同的遗传回路，即协调前馈环还存在于许多其他细菌和一些其他微生物体内。演化似乎已经因为特定模体的最优特性而选择了它们（例如，因为它们能用更少量的所需基因来执行某项功能）。此外，模体机制的明显优点是模体可以组合产生新的功能,而其中某个模体受到破坏也不会影响到别的模体。

模体乃某种小规模、局部的重复性模块。但当人们考虑社区时,他们通常意在发现网络中的大型分区,比如食物网的区划、在线社区、学科领域等等。这些结构并不表现出规律性的重复模式。如果我们掌握了某些线索,比如,假如社区的成员因某个因素而自发确定自己的身份,则相对容易找到它们。这个因素可能是社区内所有成员添加到他们博客上的小饰物、共同的着装方式等。然而,多数时候这种信息既非现成也不明确,我们必须深入挖掘网络结构以找到模块。社区识别的总体目标为发现那些内部连接比彼此之间连接更为紧密的节点集,就像空手道俱乐部网络中呈现的那样。口头表述是很容易的,但将此概念转换为数学表达却很难,以至于人们尚未发现确切的社群检测方法。一些方法可以聚合节点以满足最优性准则。其他方法则可将网络拆分为群组,然后再进一步将群组进行拆分,接着再进行拆分,进而创建一个嵌套社区的谱系树。还有一些方法在节点之间放置假想的弹簧,然后查看系统松弛之后所形成的节点群集。一般来说,还有很多其他方法上的选项。有一种有趣的技术聪明地利用了网络拓扑学,其基础在于计算**边介数**,也就是找出多数最短路径所通过的边。具有最高边介数的连接类似于在格兰诺维特的研究中连接原本相互分离之群体的弱连带。如果去掉一些高介数的边,那么,网络就会分裂成一定数量的孤立节点群集;它们便是恰当的候选社区。我们还可以继续去掉网络中一些介数较高的边,以找出嵌套在更大结构中的更为精细的结构。

社区发现方法的一个有趣应用是对美国政治博客圈的分析。物理学家拉达·阿达米克在民主党人和共和党人的博客间发现了清晰的分隔。其得出的网络结构显示，这两大党的阵营绝少相互关联。此外，民主党的博客比共和党的博客更缺乏凝聚力。例如，在此博客圈的堕胎讨论专区中，反对堕胎的博客比支持堕胎的博客联系更为紧密。因此，在线造势运动更可能在前者的博客中得到传播。另一项研究分析了美国学校中的学生社区特征，借此了解种族是否会塑造社交网络。在种族非常多元和十分同质的学校中，这一因素似乎都是无关紧要的。相反，在种族多样性处于中间值时则能看出隔离特征。在代谢网络中，人们已经发现了与特定功能相对应的社区（碳水化合物代谢，核苷酸与核酸代谢，蛋白质、肽和氨基酸代谢，脂类代谢，芳香族化合物代谢，单碳化合物代谢和辅酶代谢等）。最后，公司股票则在价格相关性的基础上聚类，人们能从中发现与银行、矿业、分销、金融等各业务领域相对应的模块。

　　将社区定义为"内部联系比外部联系更为紧密"的子图非常普遍，但这种做法并未涵盖某些特定的模块。想想朋友间的电话通信链条，其中第一个呼叫第二个，第二个呼叫第三个，以此类推：根据上述定义，这种链条极有可能不会被归类为社区。另外一个例子则是同行业竞争者的网页：显然，他们并没有动力相互连接，尽管他们明显属于同一社区。此外，真实世界的社区比密集的节点群集复杂得多。多种划分可能同时出现：国籍、社会阶层、性别、工作、政治观念统统都可用来对同一群人进行分类。而且，社区之间可能互相重叠：一个人可能同时从属于多个国籍或隶属关系。最后，嵌套社区也可能存在：例如，地域身份从属于国籍。

  尽管过于简化,但图示法仍然能够捕捉到系统的诸多相关特征。若我们仔细查看图,便会发现大量相关信息,而运行的复杂测算越多,便会呈现越多细节。真实世界的网络几乎在任何时候都会偏离其随机对照网络,这意味着其中存在某种嵌入的秩序。同样,所有这些网络都未经过设计:偏离很可能产生于自组织过程。目前,在图的结构中寻找新的规律并揭示其潜在机制是网络科学仍然面临的一些挑战。

第八章
网络中的完美风暴

意外之背景

巴罗科罗拉多岛是巴拿马运河中央的一块热带雨林。当其附近的一条河被筑坝拦截后,该岛便只剩下几处小山顶还显露在水面上。它现已成为一个露天试验场,而试验内容是高速公路、建筑物、田地或矿井替代了原始植被,雨林被分割为若干小块之后的情形。洪水淹没巴罗科罗拉多岛周边数年之后,美洲虎和美洲豹的种群数量都迅速萎缩了。结果,它们的猎物种群数量迅速攀升:现在,一种叫刺鼠的大型啮齿类动物已遍布该岛。这些啮齿动物爱好金合欢的硕大种子,所以,它们的繁盛对金合欢的成功繁殖以及依靠这些种子维生的微生物都构成了巨大难题。随着金合欢种群数量的减少,那些种子较小的植物便取而代之,而以后者为食的动物数量随之增加。生态系统的原初变化便向岛上食物网的各个方向延伸开去。

食物网中的多米诺效应并不罕见。通常,网络一般会为大规模的突发及意外动态提供背景支撑。运输系统中的病原体、电网中的断电、社会系统中的大型冲突或意想不到的合作努力等都证明了：网络似乎是"完美风暴"的理想背景。网络节点代表交换物质或信息（信息包、能量等）的单独实体（人、计算机、物种、基因……），或者它们也可以表示交换单独实体（货物、旅行者……）的场所（国家、机场……）。在这个非常宽泛的分类中，潜在动态的范围十分广大。为何网络会成为这些动态的天然发生场所？图的结构又会如何影响这些过程？对于这些问题不可能有总体答案，但在许多情况下我们能看到，底层网络的异质结构、非随机组织等特征会对表层发生的所有现象造成重要影响。

故障和攻击

2001年7月18日，一列火车在美国巴尔的摩的地下隧道中脱轨并起火。稍后，美国东海岸几个州的网速就变慢了。大火烧毁了途经隧道的光缆，这些光缆分属于美国最重要的几个互联网服务供应商。结果，这一事故引发了横扫美国大部分地区的多米诺效应。互联网常常面临类似的事故。任何时候总会有一定比例的路由器由于各种原因而一直处于死机状态，而每次类似事故都可能会与巴尔的摩脱轨事故一样严重。然而，这种大范围的破坏还是很少见的。网络似乎会容忍一定数量的长期功能障碍而不会出太多问题。这在一定程度上依赖于替代路径，后者允许网络内部流量绕过故障区域。然而，互联网与多数网络一样并没有太多冗余连接，其密度也并不是很高。考虑到这些特征，我们很自然就能预料到它比较容易发生故障。

尽管互联网似乎相对能够承受一些错误和意外，但精心设计的一次攻击仍能造成严重破坏。2000年2月7日，大量用户登录了雅虎网站，其规模之大令该公司的服务器无法响应这些请求，网页于是随之崩溃。在随后的几天里，一系列其他网页（从易趣到美国有线电视新闻网络等）也因为同样的原因而崩溃。两个月后，警察发现此类登录乃一名15岁的加拿大黑客所为，其昵称为"黑手党男孩"。他并不需要烧毁任何电缆便能阻塞互联网：其所做之事足以让这些吸引万维网上多数流量的网站崩溃。

跟互联网和万维网的情况一样，大多数真实世界的网络都显示出一种双刃的鲁棒性（robustness）。即便大部分网络遭到破坏，它们仍然能够正常运转，但某些突然的小故障或者有针对性的攻击则可能让它们彻底崩溃。例如，基因突变在整个生命过程中都会自然发生（其中有些甚至能删除细胞中的某些蛋白质），或者也能人为地产生（一项名为基因剔除的基因技术就是如此，它能关闭实验室老鼠一个完整基因的功能）。但是，生物体仍能表现出对诸多突变以及大量基因剔除的极强忍耐力。多数时候，生物体还是会在整体上继续正常工作。另一方面，某些特定的突变却能完全破坏细胞的工作。大脑一直在丢失神经元：一次给任一器官带来压力的经历，比如偶尔的酗酒，就会杀死大量细胞。但宿醉过后，一切又都恢复如初。以帕金森症为例，相当比例的神经元甚至会在病人毫无察觉的情况下消失。但当这一比例超过某一阈值，受损情况便开始变得明显。

在这方面，网络与设计而成的系统十分不同。以飞机为例，一个元件的损坏便足以让整个机器停止运转。为了让它更具复原力，我们必须采取策略，比如复制飞机的某些部件：这能让它几乎100%安全。相比之下，多数并非设计的网络则对广泛的错误表现出自然的复原力，但当特定元素失效，它们便会崩溃。网络能容纳多少错误而不出问题？而导致其崩溃的因素又是什么？为了回答这些问题，科学家通过移除网络节点以观察会发生什么情况的方式来模拟故障。删除一部分节点之后，他们会检查剩余的节点是否仍旧相互关联（即某种巨型连通分量是否仍存在于网络之中）且连接紧密（即节点之间的平均距离是否依旧很小）。为了模拟误差，节点是随机移除的。当在随机图中如此操作时，几次移除就会导致节点间距离迅速增加，图也瓦解为许多不相连部分。在半数节点移除之后，与大多数真实世界网络大小一样的随机图便遭到破坏。另一方面，异质网络（不管是真实网络图还是大小类似的无标度模型图）中经历相同的过程时，其中的巨型连接通量在80%的节点都移除之后仍然存在，而其内部的距离则实际上与最初无异。而当研究者模拟像"黑手党男孩"使用的那样一项针对性攻击时，情况则有所不同。他们一开始移除了网络中最"重要"的节点（枢纽节点）。在这种情况下，两种网络的崩溃速度都比之前快很多。然而，后者更为脆弱：在同质网络中，需要移除大约1/5的枢纽节点才能将其摧毁，而异质网络刚被移除少数枢纽节点，就会发生坍塌。

　　高度连接的节点似乎在错误和攻击中都发挥着至关重要的作用。它们是暴露在针对性攻击之下的多数异质网络的致命弱点。在这些网络中，枢纽节点主要负责图的整体聚合，移除其中少数枢纽节点便足以将其摧毁。另一方面,枢纽节点也是这些网络在暴露于错误和故障中时的"王牌":当随机移除节点时,多数时候被选出的节点均为低度数节点,因此,只要枢纽节点保持不变,网络便不会坍塌。考虑到节点度数常常与中介性相关,这种情况就会愈加清晰。高度数节点多数时候都是许多网络路径经过的桥梁。当网络遭到随机破坏,为数不多的枢纽节点很少会受到影响。既然枢纽节点不受影响,它们便提供了必要的连接;许多冗余连接则显得多余;经过枢纽节点的路径会让受损网络的工作区域保持连接状态。在某些低度数节点具备高中介性且扮演桥梁角色(就像某些机场一样)的少数网络中,枢纽节点遭到攻击仍然会导致严重的破坏,但最致命的策略还是攻击最具中心性的节点。

多米诺效应

系统可能会从能够容错的弹性状态突然转向全局性崩溃,这应该引起人们的警醒。根据估计,一定速度的物种灭绝在生态系统中是不可避免的:每年,每一百万个物种中就会有一个灭绝。通常,食物网会在灭绝事件发生之后重组,绝大多数物种并不会受到此类自然灭绝事件的重大影响。但大规模的食物网崩溃也是可能的:大约2.5亿年前,超过90%的物种在相对较短的时期里纷纷灭绝,这便是著名的二叠纪大灭绝事件。过去5亿年里,地球上总共发生过5次类似的大灭绝事件。研究人员认为地外因素可能是罪魁祸首,比如充满争议的可能导致恐龙灭绝的陨石坠落事件。然而,也可以用网络来解释这些事件。物种的连环灭绝或共同灭绝事件对于生态学家而言并不陌生。例如，英国曾于20世纪中期引入黏液瘤病毒以控制兔子的种群数量，这最终导致大蓝蝶于1979年灭绝。这种病毒消灭了兔子，但兔子食用的高茎草却因此得以蔓延。后者又破坏了蚂蚁的栖息地，它们习惯在有阳光的低茎草中筑巢。蚂蚁与大蓝蝶的幼虫之间有着共生关系：它们会照看幼虫，而幼虫则报之以流食。因此，蚂蚁栖息地的破坏逐渐损害了大蓝蝶的繁衍，进而导致其灭绝。这并非真正意义上的共同灭绝，因为兔子并没有因为黏液瘤病毒而灭绝，而部分大蓝蝶也被重新引进。然而，这一事件却让我们对食物网可能受损的程度有所了解。将此事件的规模扩大，其中的连环灭绝几乎耗尽整个生态系统的物种，这便可能成为以往大灭绝事件的另一个解释。当人类主动对生态系统进行大规模破坏时也应考虑这种情况，例如，目前过多的捕捞正以前所未有的规模消耗着海洋生态系统。

　　网络上少数其他动态过程也可能引发类似的**连锁故障**或**雪崩式崩溃**。大规模停电便是一个典型例证：某个发电站的故障导致另一个发电站过载，这又导致后者出现故障，继而在网络中大范围传播电力过载故障。在此现象中，某一节点的故障不仅会导致节点互联的损失或降低节点之间的平均距离，而且还会引发多米诺效应。金融危机中出现的经济网络的**系统故障**则是这种现象的又一例证。**拥堵现象**中也会发生同样的情况，比如车流致使街道网络某些地点的交通发生瘫痪，人流因特殊事件而在地铁站中形成阻塞，又或者网络流量导致某些互联网服务器崩溃，等等。研究表明，枢纽节点在所有这些情况中都至关重要，不仅因为它们能减少运输时间，还因为它们会率先饱和。

流行病

1347年,一场人类史上最具破坏性的瘟疫在君士坦丁堡暴发。之后的三年里,黑死病蔓延至欧洲,导致该地区大部分人口死亡。这种疾病像波浪一样以每年200—400英里的速度在欧洲蔓延开去(图12左)。这种传染图景与现代流行病大异其趣。据估计导致世界3%人口死亡的1918年大流感仅用了一年时间传播,它甚至蔓延到了与世隔绝的太平洋岛屿上。1957年的"亚洲流感"病毒仅用了半年时间便席卷全球。而更近的2009年的猪流感则在几星期之内便从地球一侧蔓延到了另一侧(图12右)。黑死病潜伏在船只和车厢中经由朝圣者、商人和水手等媒介传播的速度每天不过数英里,而现代疾病则依赖更高效的交通方式传播,比如高速公路、火车和飞机等。14世纪时,物理距离是流行病传播的主导因素。而在现代的网络世界中,传染病能跳上飞机,数小时之内便可到达地球的另一端。

流行病经由全球网络(比如机场网络)和地方网络传播;人际传染病则依靠个人社交网络传播。例如,流感在一定程度上通过个人之间的面对面接触传播,艾滋病病毒则经由无保护性接触网络传播。2001年,卡比兰公司的物理学家罗慕阿尔多·帕斯托尔·萨托拉斯及其意大利同事亚历山德罗·韦斯皮尼亚尼决定通过建模和模拟疾病在社交网络中的传播来研究这个问题。他们只引入了疾病传染过程的最简机制:一开始,社交网络中的少数个体感染了疾病;如果某健康个体通过某种关联与这些人中之一有过接触,则他或她有一定的概率会被感染;另一方面,被感染个体也有一定的概率会康复。这种感染

图12 14世纪的瘟疫像波浪一样在欧洲蔓延(左),而2009年的猪流感则更像火团从地球的一端至另一端溅下火花(右):其中的差异在于人类交通网络的巨大变化

模型被称为"SIS",因为每个人都会经历"易感–感染–易感"(susceptible-infected-susceptible)这个周期(健康个体对疾病"易感")。这个过程表现了类似普通感冒的传染病,感染人群往往会在这个过程中恢复。还可以将这个过程复杂化以表现其他传染病,例如引入致死或免疫等可能性。然而,这些修改无法改变最终结果的大致走向。

萨托拉斯和韦斯皮尼亚尼发现,病毒在最初的扩张之后要么被根除——迅速减少并最终从人群中消失——要么成为地方病——停留在某地并反复感染该地区的部分人群。若每个被感染个体所感染的人数少于一人,则该种疾病低于**传播阈值**:在这种情况下,该疾病会逐渐消失。若每个被感染个体所传染的人数超过一人,则该疾病便已超过传播阈值:在这种情况下,该疾病会逐渐传播开去。若疫苗可用,人们便可通过使足够比例的人口免疫而将相关疾病控制在传播阈值之下。高传染性疾病常常最为棘手,因为它们的传播阈值很低,进而十分容易传开去。如果人们难以根除某种疾病,则将其传染性降至传播阈值附近仍有积极效果:这会降低受地方性疾病反复感染的人口比例。

在萨托拉斯和韦斯皮尼亚尼的研究中,他们发现传播阈值主要取决于底层网络的特征。当SIS机制作用于随机网络时,出现了一种能够使我们估测根除相关疾病所需免疫人数的明确阈值。但当这种机制作用于异质网络时,免疫阈值则几乎消失:它远低于随机图的阈值;此外,系统规模越大,免疫阈值越低。若网络足够大,免疫阈值会低至几乎无法避免部分人群被感染的程度。不将几乎全部人群进行免疫是无法将疾病控制在如此之低的阈值下的。流行病与其他许多动力机制一样,都会受网络异质性的影响。人们对故障和攻击的研究已经表明,枢纽节点会将网络的不同部分连接。这意味着它们也扮演着传播疾病之桥梁的角色。枢纽节点的众多联系将它们与被感个体和健康个体相互联系:因此,枢纽节点很容易被感染,它们也容易感染其他节点。流行病学家确认的**超级传播者**很可能就是社交网络的枢纽节点。

异质网络在流行病面前十分脆弱不是个好消息,但加深对它的理解却会为疾病控制提供好的思路。理想情况下,几乎所有群体都应被免疫以完全阻断传染。然而,如果我们只能免疫一小部分人群,随机选择免疫人群便不是个好主意。多数时候,随机选择意味着会选出那些与他人联系相对较少的个体。即便这样能让他们阻断疾病在其周围的传播,但枢纽人物还是会让疾病重新传播开去。瞄准枢纽人物是更好的策略。对枢纽人物进行免疫就像将其从网络中删除,而对针对性攻击的研究表明,删除小部分枢纽节点会打碎网络:因此,疾病将被限于网络中少数孤立区域。这个策略面临着一个实际问题:无人真正知晓某个人群的完整社交联络图,所以很难确定枢纽人物。然而,物理学家鲁文·科恩、什洛莫·哈夫林以及丹尼尔·本-亚伯拉罕在2003年提出了一个精巧的策略以找到枢纽人物:他们建议对人群随机抽样,并询问与这些被选取人相互联系的人的名字。这个名单中重复最多的名字最可能是社交网络的枢纽:事实上,枢纽人物因自身的众多连接而与很多人相关联,所以许多受访者都会提到他们。我们应当注意,免疫枢纽人物在理论上十分奏效,但真实世界的诸多细节会降低其有效性,比如,网络是否会处理枢纽节点周围的特殊冗余路径，或者联系网是稳定不变还是在不断演化等：例如，若爱丽丝携带艾滋病病毒且与鲍勃有过无保护性行为，而鲍勃又与卡罗尔有过无保护性行为，这两次性行为发生的先后顺序对卡罗尔而言则意义重大。

流行病在社交网络中的传播图景可部分推广至其中节点不表示人群而代表地点（比如机场）的网络中，并且在这个网络上传播的是人群（比如被感染或健康的旅行者）。在这种情况下，我们可以定义一个**全球入侵阈值**，高于此阈值的疾病成为流行病，反之则为局部层面的疾病。关闭机场往往并非上策：我们需要关闭90%的机场才能有效阻止某些流行病的传播，这将带来巨大的社会和经济损失。与发展中国家（这些国家通常是新型流行病的源头）共享抗病毒药物这种更聪明的策略则有效得多。

病毒、广告与时尚

两个无名的巴基斯坦程序员，一位大学教授，一群高中生……这些人便是计算机病毒的始作俑者。上世纪80年代，这些寄生病毒程序开始在计算机之间传播，它们基本上藏身于用户互换的软盘上。第一批病毒是自我复制软件的学术实验品，但它们很快从实验室中逃逸。1986年，"大脑"病毒出现在巴基斯坦。同年，一家德国实验室失去了对"波光"病毒的控制。一年后，一群学生传播了"维也纳"病毒。90年代，计算机病毒已成为全球性问题，但这与即将发生的事情相比则微不足道。

互联网的出现也带来了新一代病毒，它们能通过网络将自己拷贝发送至其他计算机。1999年，"梅丽莎"病毒在互联网蔓延：人们开始收到主题为"有你的重要消息"或"这是你请求的文件……请勿转给他人；-)"的邮件信息。这些邮件中包含一个名为"list.doc"的文件。如果接收者将其打开，它就会启动一个程序，进而将相同的信息发送至计算机中保存的前50位联系人。"爱虫""蓝宝石""巨无霸""冲击波"以及其他许多计算机病毒纷纷在网络上爆发，它们有着类似的机制并都造成了灾难性后果：其中一些破坏了公司的计算机系统、大学的数据库，甚至影响了互联网流量。

计算机病毒传播中的一些特征与真实世界的流行病惊人地相似。计算机经由自己的连接（例如，作为计算机主人社交联系人样本的电邮网络）而受到感染，同时也以类似的方式感染其他计算机。一些关于疾病的结论解释了计算机病毒令人困惑的行为。即便杀毒程序会及时更新，一些病毒仍会在其首次攻击后的数年里继续传播。若考虑到流行病在无标度网络中的流行特征，这便不足为奇了：即使大部分计算机因杀毒程序而免疫，也不足以根除病毒感染：总会在这里或那里存在一些高度数节点让病毒重新流行。

若想在异质网络中传播信息，对计算机病毒构成严重问题的异质网络容错特性则可转变为某种资源。这便是**病毒式营销**背后的原理。多亏了虚拟社交网络，万维网如今充满了"像病毒般扩散"的视频、游戏和应用程序：每天都会有成千上万人将这些内容转发给自己的所有联系人。体现这个想法的首个例子是微软电邮服务的传播。1996年，微软公司在邮件中插入写有"来微软电邮获取你的免费网页邮箱"的自动页脚，其中包含的链接能让人在很短的时间里设置一个新邮件地址（见本书第50

页）。类似的策略也体现在雅虎、谷歌的邮箱服务以及许多基于邀请才能使用的社交网络服务中。

病毒式营销利用了一种被称为**社交传播**的潜在心理现象。这是人们模仿其联系人传播闲话、时尚、谣言和观念的大体趋势。这种心理机制也会在创新、团队解决问题和集体决策中起作用。社会学家和心理学家们发现了人类带有明显彼此"模仿"倾向的很多例子。1962年，坦桑尼亚一所教会学校中的一群女孩就经历了一场无法控制笑声的非正常倾向。数月之后，同一所学校里的几十名学生也都表现出相同的症状，而这些学生中的一部分被送往一些村庄安置后，村中其他人也表现出同样令人不安的咯咯笑的症状。经过大量调查，研究该病例的医生A.M.兰金和P.J.菲利普得出结论：这是"集体歇斯底里"的一个病例。1998年田纳西州的一所高中发现了一个类似案例，当时，感觉闻到汽油的老师将这种感受传递给了数百名学生。排除所有外部环境因素后，科学家们得出结论说，某种"情绪传染"机制在其中起了作用。

人们已记录了许多类似的社交传播案例，但近年来科学家们发现，同样的机制还可能在不那么特殊的环境中起作用：例如，肥胖和吸烟似乎也会在社交网络中传播。相互关联的人群共有某种特征或行为主要基于三个原因。首先在于他们属于同一社会阶层这种外部因素：例如，同属较低社会阶层的人吸烟和肥胖的风险较高；同时，与社会阶层较高的人相比，他们更有可能与彼此建立联系。其次，人以群分：吸烟者或那些有着相似体重的人们往往会与有着类似癖好的人成为朋友。第三，社交传播的作用：如果你是吸烟者或超重者的朋友，那你更有可能开始吸烟或增加自己的日常食物摄入量。这三种因素可能同时起作用，但社交传播可能最为重要，不应被低估。社会学家认为，传播的并非某种特定的情况；相反，是人们对何谓恰当的共识得到了传播。这种视角可应用于公共卫生领域，通过采取针对社交网络枢纽节点的措施来培养人们更加健康的习惯。

自然，行为、谣言及观念的传染在许多方面都与疾病传播有所不同。与传染病不同，散布信息必然是有意为之。另一方面，获取信息往往对人有利，因此它更多地是一种主动而非被动感染的过程。学习或被说服所需的接触时间可能比患上疾病要长。此外，许多其他竞争因素也会起作用。如果社交传播是主导因素，均一性将是其内在规则，但事实上，反对简单同化的因素会产生多样性、少数群体和两极分化。无论如何，社交传播在一定背景下或许的确是最为相关的因素。理查德·费曼于1940年代发明了**费曼图**这种现代高能物理学工具。一些物理学家满心热诚地接受了这种工具，另一些人则对其抱有疑虑，但这些图最终成功了。针对费曼图在美国、日本、苏联物理学家团体中扩散的研究表明，其扩散趋势可与流行病模型十分准确地吻合，前提是将模型中的参数调整为十分不同的值。

网络及动力机制孰先孰后？

古罗马成功的关键因素之一是其紧邻台伯河的战略地位，该河当时乃首屈一指的通信和商业路线。城邦变得更加强大之后，人们便开始营建其强大的道路网络分支。反过来，道路网又成为维持并进一步扩张罗马强权的关键工具，因为路网使得运输货物和军团更加快捷。更多的道路意味着更多的权力，而更多的权力又必然会打造更多道路：结果便如意大利谚语所言，"条条大路通罗马"。我们几乎能在每个重要城市的历史上发现类似的发展模式。扩张中的城市吸引了更多的交通，也需要更多的连接方式（道路、铁路、航线……），这些因素反过来又会增加交通流量和扩大城市规模，而这又意味着更多的连接。通信网络会影响交通的动力，后者反过来又会在反馈回路中重新塑造网络。

网络拓扑如何影响了动力机制这一问题暗含了某种假设：网络乃不可变的结构，过程在它上面发生。现实中，所有网络都会在动力作用的**过程**中发生改变。因此，仅当动力机制的发生时间比拓扑早很多时，这一假设才有意义。在某些过程中这是合理的：例如，人们每天或每周相互交换的信息会在固定的社交网络中传播，因为通常友谊与亲属关系的更替是按年进行的；或者，城市的车辆交通不论在哪一天都会在固定街道上进行：通常街道连接不会每天都改变。

然而，在其他情况下，这种假设是有缺陷的。例如，在性传播疾病的传播过程中，与他人发生关系的时间顺序尤为重要。与某人建立无保护性关系发生在与受感染的他人建立无保护性关系之前还是之后很不一样。如果我们想研究一个城市十年的发展，那么有必要考虑交通和正在改变的连接之间的相互作用。在类似对等文件共享系统等一些技术网络中，网络结构和信息动力机制在相同的时间段内变化且强烈地相互交织。而食物网中的种群动态能够引起网络的重组。当过度捕捞将某物种数量降至一定水平之下时，食物网会重新排列捕食次序，并让新物种代替旧物种。网络结构和动力机制的耦合在虚拟社交网络发展的特定时刻尤其重要。这些工具提供了个人网络结构和内容的恒定信息流。因此，研究者认为，这种增强的意识可能改变人们创作、维护和影响其社交网络的方式。

　　一些方法可用于处理网络结构和动力机制耦合这个问题。例如，我们可以通过最优化构建网络模型，其中需要优化的量与网络流量或搜索这样的动力机制相关。更加完善的方法在于修改适应性模型，使得适应性的值取决于一些动态参数。当动力机制继续作用时，适应性也会相应地发生改变；这便可以重组网络。其他策略也是可行的，所有这些都巩固了一个基本观点：多数情况下，当某个动力机制在网络结构中发生或与它耦合时，多数时候我们必须将基本图纳入考虑以充分理解正在发生的情形。

第九章

整个世界是否就是一张网？

据报道，量子力学的创始人之一保罗·狄拉克在谈及物理学家们在20世纪初的革命性发现时说："其余都是化学过程。"他的意思是，所有科学都能从物理学的基本原理中推导出来。很不幸，仅有少数几种情况可以由量子力学方程精确求解，它们还基本都是氢氦原子。像分子那样更为复杂的东西则必须以近似法或计算机模拟的方式处理。目前，除了少数宏观量子效应，基础物理学当前对我们理解生物、心理或社会而言相对无用。类似的错误还发生在遗传学中，人们错误地将DNA界定为能够解释人类所有特征、疾病和行为的决定因素。一般而言，基础科学成果的应用不应超出其真实的有效范围，人们应当认识到更专业的学科能够提供超出这一范围的更深刻洞见。网络科学也应避免落入被夸大的圈套。其整体性的视野，对差异明显的系统出人意料的相似性的揭示，还有当前对网络概念的文化上的迷恋，所有这一切使得我们很容易将网络科学视为"万用理论"。社会学家、工程师、生物学家和哲学家都曾警告世人注意从网络理论中提取的空洞概括。这类批评多数是合理的,但网络科学的成果不应被低估,其对未来各种发现的潜力也不应受到贬低。

网络科学的第一个主要限制是其缺乏大规模数据。社会科学中使用的调查问卷和访谈等方法十分昂贵且耗时,有时候还容易带有主观偏见。从信息技术(通话记录、电子邮件、社交网络、地理定位、无线射频识别芯片、健康数据、信用卡等等)处获取的数据为我们了解人们的社交关系提供了空前的便利,但这也带来一些问题。披萨送货员会接到许多电话,但多数来自客户而非其朋友:"披萨送货员问题"表明,人们从信息技术数据(比如本例中的通话记录)中整理出相关信息并不容易。此外,还应提及的是,网络数据挖掘也制造了一些涉及隐私以及军事用途的伦理问题。

很多情况下,人们仅能获取部分信息。为了画出水生生物的食物网,生态学家会捕捉鱼类并检查其消化道:即便最杰出的科学家用这种方法也会错过一些物种联系。推导出蛋白质之间物质相互作用的基因方法既能产生假阳性也能产生假阴性结果。互联网和万维网图可以通过从某节点释放"探测器"以探索其周围各边的方式获取:足够数量的路径能很好地呈现网络,但某些边可能从不会被探测器发现。

一旦数据可用,将其呈现在图中则不可避免会将其简化。网络方法的优势之一是其对拓扑的关注,但它忽略了元素的诸多具体特征。如果我们对这些特征感兴趣,图形近似法便显得不够用了。网络模型有时候被修改后可以包含这些特征,但并不总是如此。

当地理（即节点的物理位置）的重要性高过拓扑时，图示法也会不够用。例如，变电站、机场或火车站的位置明显与其连接的安排有关。此外，社交网络和食物网中的节点接近性则能决定建立某种关系的实际可能性。图示法可能遗漏的另一个元素是时间。例如，在性传播疾病中，在某人被他人感染**之后**与其建立关系明显不同于在这**之前**与其建立关系：关系建立的时机对疾病的传播至关重要。时机在大量网络中都很重要：例如，科学出版物按时间排序，科学家们只能引用那些之前发表的文献。

有时，识别节点和边绝非小事。我们很容易区分鹰和隼，却很难数得清生态系统中的细菌数量。为了避免低估"微小"物种的数量，生态学家通常会将**营养物种**里的生物聚集，这些生物共享同样的天敌以及相同的猎物。类似地，社会科学家则会聚集**结构上等同**的个人：具有相同联结数量和种类的人，比如家庭成员。类似的方法还被用于自主系统层面的互联网以及大脑区域网络等。这些过程必须以连贯的方式进行以获得一个合理的网络。边的定义更为复杂。某公司可能持有另一个公司的小部分或全部资本。两座机场可通过每天一次航班或每小时一次航班相互连接。在所有这些情况中都必须建立阈值，低于该阈值则被视为弱关系而不被记录。为连接加权或设置阈值会强烈影响所得网络的形状，因此必须建立在充足的理由上。

一旦将数据组织为图，就必须仔细解读其结果。网络科学有一整个学科分支都在专门研究可视化，即制造算法以便在纸张或计算机屏幕上合理安排节点和连接。然而，多数结论无法经由目测得出，而要对其进行数学分析。一些批评指出，并非所有复杂网络都具备异质度数分布（在任何情况下，这都绝非数学上的精确幂律)。的确,若网络并非异质,它们也会有趣,但公平地说,有趣的网络往往是异质的。完美的幂律并不重要:重要的是肥尾的出现,它揭示了枢纽节点的存在。将异质性解读为自组织的标记这一点已受到批评,人们指出许多网络中都存在着一定程度的规划,正如管理员对局域网的具体设计一样。但无疑地,互联网与其他许多网络一样并未经过大规模设计。因此,人们有理由认为,这些网络对随机性的偏离可被归因于自组织过程。此外,度数异质性仅是网络复杂性的其中一个标记。异质性还出现在许多其他特征中,比如中介性、集聚性和权重等,复杂性同样也会出现在其他特征中,这一点与异质性不同:例如,模块和社区结构常常偏离随机性,并对网络顶层的动力机制产生强烈影响。

另外一个批评则指出,网络科学发现的只是不同系统和动力机制之间的模糊相似性,而非真实的**普适类**。后者是对应于相同基本数学定律(在特定细节被忽略的情况下)的不同现象群。当然,生物网络的具体特征与技术网络的具体特征完全不同,计算机蠕虫病毒的扩散与疾病的传染遵循着不同的规律。然而,网络理论为这些如此不同的结构和过程提供了一个共同的趋势和预测的框架。通常,如果系统足够大且相关现象的观察时间足够长,它们便会展现出十分相似的趋势。

网络科学已经在一些领域显示出了它的预测能力。它目前被应用于咨询行业,以帮助组织更好地开发各成员不同的技能;在公共卫生领域,它能预测和防止传染病的蔓延;在警察和军事系统中,它被用于追踪恐怖分子、罪犯以及反叛团伙;此外网络科学在其他一些领域也有应用。还有许多问题亟待解决,其中包括：制作更精细的模型以适配特定的网络和动力机制；寻找新的相关网络数据；深入挖掘拓扑以发现未被注意到的规律，并充分解释现有规律；描绘小型网络的特征并学习如何处理众网之网；将生物网络与演化范式更有效地连接；发现新的应用（比如在药物设计中）；以及有可能的话找到普适类。

伽利略·加利莱伊有一句名言如此说道："哲学写在宇宙这本大书之中……书写语言为数学，字母则为三角形、圆形以及其他几何图形……"我们相信，尤其在如今这个复杂的当代世界里，我们需要网络这种"字母"。

索 引

（条目后的数字为原书页码，见本书边码）

A

Achilles' heel 致命弱点 97
Adamic, L. 拉达·阿达米克 92
agouti 刺鼠 94
Albert, R. 雷卡·奥尔贝特 70；参见 Barabási-Albert model
algal blooms 藻类暴发 27
'All roads lead to Rome' "条条大路通罗马" 108
Anchorage, Alaska 阿拉斯加州安克雷奇 88
anchovy fishery collapse, Peru 秘鲁鳀鱼业崩溃 28
Ansell, C. K. 克里斯托弗·K. 安塞尔 83
Antal, T. 蒂博尔·安塔尔 21
ants 蚂蚁 2, 99
Argentina 阿根廷 17
Arpanet 阿帕网 37—38
Arunda people 阿伦达人 10
Asian flu 亚洲流感 100
assortative mixing 相称混合 81—82
ATP molecule 腺苷三磷酸分子 56
autonomous systems 自治系统 39, 82, 85, 88, 112
average degree 平均度数 45
Avraham, D. ben 丹尼尔·本-亚伯拉罕 103

B

Bacon, Kevin 凯文·贝肯 46—47
Baltimore (US), derailment 巴尔的摩（美国）脱轨事件 95—96
Barabási, A.-L. 奥尔贝特-拉斯洛·巴拉巴西 70
Barabási-Albert model 巴拉巴西-奥尔贝特模型 71, 74—77
Baran, P. 保罗·巴兰 38
Barro Colorado island 巴罗科罗拉多岛 94
basal species 基位物种 28, 53
Beck, H. 亨利·贝克 14, 16
bell curve 钟形曲线 60—16
benefit-based antecedents 基于利益的前因 79
Bernard of Chartres 沙特尔的伯纳德 34
Berners-Lee, T. 蒂姆·伯纳斯-李 39
betweenness centrality 中介中心性 86—87
Bianconi, G. 吉内斯特拉·比安科尼 77—88
big blue butterfly (*Maculineaarion*) 大蓝蝶 98
bipartite cliques 二分团 90
Black Death 黑死病 100—101
blackouts, electrical 停电 5, 42, 99
blogospheres 博客圈 92
bottom-up process 自下而上的过程 64
bow-tie structure "蝴蝶结"结构 44;

115

参见 directed graphs
Bozon, M. 米歇尔·博宗 78
Brain virus "大脑"病毒 104
breakdown avalanches 雪崩式崩溃 99
Broca, P. 保罗·布罗卡 25
Broca's area 布罗卡区域 26

C

C. elegans (nematode worm) 秀丽隐杆线虫(线虫) 23, 26, 49, 62
CAIDA (Cooperative Association for Internet Data Analysis) 互联网数据分析合作协会 66
Cailliau, R. 罗伯特·卡约 39
Caldarelli, G. 圭多·卡尔达雷利 76
Capocci, A. 安德里亚·卡波奇 76
cascading failures 连锁故障 99
Caulobacter crescentus 新月柄杆菌 56
Cayley, A. A. 凯利 9
centrality 中心性 86—88, 90
cerebellum 小脑 26
chains 链条 2, 17—19, 47—48, 86, 92—93
characteristic scale 特征尺度 58—59, 63
Christakis, N. 尼古拉斯·克里斯塔基斯 85—86
circles of women 妇女群体 10
citations networks 引用网络 34, 68, 69—70
cliques 团体 90
closeness centrality 接近中心性 87
clustering 集聚性 84—85, 90, 113
clustering coefficient 集聚系数 84
co-extinctions 共同灭绝 29

cocktail party experience 鸡尾酒会经历 47
cod population collapse 鳕鱼种群崩溃 1—3
coenzyme A 辅酶A 74
Cohen, R. 鲁文·科恩 103
coherent feed-forward loop 协调前馈环 91
Colorado Springs 科罗拉多州斯普林斯市 10
community 社区 82, 91—93, 113
compartments 区划 14, 90—91
complex systems 复杂系统 2, 12, 17—18
computer viruses 计算机病毒 5—6, 49, 104—105
congestion phenomena 拥堵现象 99
connected triples 连接三元组 84
contagion 传染 80, 86, 106—107, 114
copying mechanism 复制机制 73—74
core discussion network 核心讨论网络 19
core vs. periphery 核心与外围群体 80
Creutzfeldt-Jacob disease (mad cow syndrome) 克雅二氏病(疯牛综合征) 24
CtrA CtrA基因 56
cut-off 上限 61—62
cyberspace 赛博空间 40—41

D

Dangi, Chandra Bahadur 钱德拉·巴哈杜尔·唐吉 58
data mining 数据挖掘 12—13, 111

Davis, A. A. 戴维斯 10
degree distribution of graphs 图的度数分布 60
dense networks 紧密网络 45, 53, 59
dichotomous connections 二分关系 21
Dirac, P. 保罗·狄拉克 110
directed graphs 有向图 21; 参见 bow-tie Structure
disassortative mixing 不相称混合 81—82
diversification 多样化 73—74
divide et impera strategy 分而治之策略 83
DNA (deoxyribonucleic acid) DNA（脱氧核糖核酸）23—24, 73—74, 110
domino effects 多米诺效应 1, 28, 94, 95, 98—99
Dunbar number 邓巴数字 62
duplication 复制 73—74
dyadic relations 二元关系 86
dynamics 动力机制 14, 29, 66, 68, 76—77, 79, 86, 94—95, 102—103, 108—109, 113—114

E

E. coli 大肠杆菌 90—91
East Asia 东亚 87
edge betweenness 边介数 92
edges 边 9, 13, 18—19, 21, 24, 27, 32—33, 36, 39, 43, 45—46, 64, 81—82, 84, 90, 111—112
educated guess 有根据的推测 12—13
Ego 埃戈 30
ego networks 自我中心网络 83

electric grid accident 电网事故 5
electrical blackouts 停电 5, 42, 99
electrical circuits 电路 9
electronic traffic 电子信息流 12, 96, 99
emergence of networks 网络的涌现 66—79
emergent phenomena 涌现现象 2, 4
emotional contagion 情绪传染 106
epidemic threshold 传播阈值 102
epidemics 流行病 100—104, 107
Erdös, P. 保罗·埃尔德什 11, 45, 46—47, 50, 51, 56
exclusion mechanism 排斥机制 18
expressive aphasia 表达性失语症 25
extinction 灭绝 98—99

F

factory workers 工人 10
fat tails 肥尾 59—63, 65, 70, 113
FC Barcelona, Messi's performance with 梅西在巴塞罗那足球俱乐部队中的表现 17
Feynman, R. 理查德·费曼 107
financial crisis, 2008 2008 年金融危机 35—36
fit-get-richer mechanism "适者更富"机制 77
fitness of a node 节点适应度 76
fluctuations 波动 59
food chains 食物链 2, 17, 28
foodwebs 食物网 29, 42—44, 53, 56, 62, 67, 82, 86, 90—91, 94, 98—99, 109, 111—112
football 足球 4, 17, 19

Fowler, J. 詹姆斯·福勒 85—86
Framingham, Massachusetts 马萨诸塞州弗雷明汉 85—86
Freeman, L. C. 林顿·C. 弗里曼 86
friendship networks 友谊网络 10—12, 20—21, 30, 58—62, 80—86, 89
functional magnetic resonance 功能性磁共振 27

G

Galileo Galilei·伽利略·加利莱伊 114
Gall, F. J. 弗朗茨·约瑟夫·加尔 25
Gardner, B. B. B.B. 加德纳 10
Gardner, M. R. M.R. 加德纳 10
Garlaschelli, D. 迭戈·加拉斯凯利 77
gene knockout 基因剔除 96
gene regulation 基因调控 24
giant connected component 巨型连通分量 43—45, 97
giant strongly connected component 巨型强连通分量 43—44
Giovanni, Don 唐·乔万尼 55
girl (or boy) next door 邻家女孩（男孩）78
global invasion threshold 全球入侵阈值 104
glycolysis 糖酵解 74
Goliath 歌利亚 56—57
Granovetter, M. 马克·格兰诺维特 30—31, 53, 92
graphs 图 4—5, 8—9, 19, 21—22, 45, 70, 81, 90, 93, 95, 111—113

chain of nodes as 节点链 17, 59, 83—84
degree distribution 度数分布 60
vs. lattices 与格子框架 18
maps as 地图 16
random 随机图 11—12, 45—46, 51, 63, 76, 81, 85, 97
subgraphs 子图 92
grids 网格 17—19, 51
group approach vs. individuals approach 群体方法与个体方法 13—14
GTP 鸟苷三磷酸 74
Guare, J. 约翰·瓜雷 48

H

Hamilton, W. R. W. R. 汉密尔顿 9
happiness 幸福 85—86
Havlin, Shlomo 什洛莫·哈夫林 103
height 身高 56—61
Henry VI, King《亨利六世》32
Heran, F. 弗朗索瓦·埃朗 78
heterogeneity 异质性 58—65, 71, 87, 97, 102—103, 105, 113
hidden variable of a node 节点的隐变量 76
homogamy 同配生殖 75
homogeneity 同质性 14, 57—61, 63—64, 71, 97
homophily 同质相吸 75, 77, 81—82, 107
HP 惠普公司 30
hub policy 枢纽政策 72
hubs 枢纽 54—56, 58—59, 61—64, 72—74, 85,

97—98, 103—104

Hudson School for Girls, New York State 纽约州的哈德森女子学校 9—10

human genome sequencing 人类基因组测序 23

hyperdiadic spread 超二元扩散 86

I

Iloveyou virus "爱虫"病毒 49, 105

immunization 免疫 102—105

in-degree 入度 21, 63

indirect mate choice 间接择偶 72

individuals approach vs. group approach 个体方法与群体方法 13—14

influenza 流感 100—101

interbank network 银行同业网络 35, 67

interlocking 连锁 35, 43, 72

intermediate species 中位物种 28

internet 互联网 4, 5, 16, 37—40, 42, 45, 49, 51, 56, 64, 66, 72—73, 78, 85, 95—96, 105, 111—113

internet dynamics 互联网动态 66—67

internet mapping projects 互联网绘图计划 66, 73

Internet Service Providers (ISPs) 互联网服务提供商 39, 72—73

internetworking 网际互连 38

isomers of organic components 有机成分异构体 9

J

Jennings, H. 海伦·詹宁斯 10

K

Kirchhoff, G. R. G.R. 基尔霍夫 9

Kochen, M. 曼弗雷德·科亨 48

Königsberg bridges 哥尼斯堡七桥 7—9

Krapivsky, P. 保罗·克拉皮夫斯基 21

Krebs, V. 瓦尔迪斯·克雷布斯 55

L

large-scale data 大规模数据 52, 111

lattices 晶格 18, 26, 51—52

laughter epidemics 笑声传染病 106

Laumann, E. 爱德华·O. 劳曼 80, 82

Loffredo, M. 马里耶拉·罗弗雷多 77

London Underground 伦敦地铁 14—16

Los Rios, P. de 保罗·德·罗斯·里奥斯 76

M

Maculinea arion (big blue butterfly) 大蓝蝶 98

mad cow syndrome (Creutzfeldt-Jacob disease) 疯牛综合征(克雅二氏病) 24

Mafia Boy "黑手党男孩" 96—97

map colouring problem 地图着色问题 9
maps of language 语言地图 33
Marsden, Peter 彼得·马斯登 19
Matthew effect 马太效应 69
Mayo, E. E. 梅奥 10
Medici, C. de' 科西莫·德·美第奇 82—83, 86
meet-me-rooms 汇接机房 56
Melissa virus "梅丽莎"病毒 105
Mercator "墨卡托"计划 66
Merton, R. 罗伯特·默顿 69
Messi, Lionel 里奥内尔·梅西 17
metabolic hubs 新陈代谢枢纽 74
metabolic network 新陈代谢网络 25, 41, 56, 59, 74, 92
metabolic pathways 新陈代谢路径 24, 67, 74
metabolism 新陈代谢 24, 92
metrics vs. topology 度量与拓扑 16
Milgram, S. 斯坦利·米尔格拉姆 47—49, 54—55, 61
modules 模块 26, 89—92
Moreno, J. 雅各布·莫雷诺 10
Moscow 莫斯科 87
motifs 模体 90—91
Mr Jacobs 雅各布先生 54—55, 61
multiplexity 关系的多重性 19
multiplicative noise 乘性噪声 72
Muñoz, M. Á. 米格尔·安琪儿·穆诺兹 76
mutualism 共生 19
myxomatosis 黏液瘤 98

网络

N

NAD 烟酰胺腺嘌呤二核苷酸 74
nematode worm (*C. elegans*) 线虫(秀丽隐杆线虫) 23, 26, 49, 62
neocortex 新皮质 26, 50, 51
network dynamics 网络动力机制 66—68
networks of networks 众网之网 38, 114

O

'-omics' revolution "组学"革命 25
open architecture 开放体系结构 38
opportunity-based antecedents 基于机遇的前因 79
optimization 最优化 79, 109
Oracle of Kevin Bacon "凯文·贝肯指南" 46—47
out-degree 出度 21, 63

P

Padgett, J. F. 约翰·F. 帕吉特 83
Pareto, V. 维尔弗雷多·帕累托 65
Pastor-Satorras, R. 罗慕阿尔多·帕斯托尔·萨托拉斯 100, 102
periphery vs. core 外围与核心 80
Permian extinction 二叠纪大灭绝 98
Philip, P. J. P. J. 菲利普 106
pizza delivery guy problem "披萨送货员问题" 62, 111
popularity is attractive principle "受欢

迎具有吸引力"原则 72

power grids 电网 1, 37, 40, 42, 62, 67, 94

power laws 幂律 60—62, 65, 70, 113

degree distributions 度数分布 71, 78—79

predation 捕食 13, 19, 22, 28—29, 53, 109

preferential attachment 优先连接 70—75, 77—79

prions 朊病毒 24

professional collaboration networks 专业协作网络 31, 62

protein interaction networks 蛋白质相互作用网络 24, 41, 56, 74

R

racial inequality 种族不平等 80

random graphs 随机图 11—12, 45—46, 51, 63, 76, 81, 85, 97

randomization 随机化 64

Rankin, A. M. A. M. 兰金 106

Rapoport, A. 阿纳托尔·拉波波特 10, 11

reach centrality 抵达中心性 87

Redner, S. 西德尼·雷德纳 21

Rényi, A. 阿尔弗雷德·雷尼 11, 45, 51

rich-get-richer mechanism *see* preferential attachment "富者更富"机制，见 preferential attachment

Rothenberg, R. B. et al 理查德·B. 罗滕伯格等人 10

routers 路由器 38—39, 45, 49, 51, 56, 66, 73, 95

S

Salomonoff, R. 雷·萨洛莫诺夫 11

self-organization 自组织 5—6, 63—65, 67, 68, 77, 79, 81, 84, 93, 113

sexual interaction networks 性关系网络 31, 43, 50, 55, 75, 80—81, 100

sexually transmitted diseases 性传播疾病 31, 50, 55, 80, 82, 108, 112

shareholding network 持股网络 35—36

shortcuts 捷径 51—53

Simon, H. A. 赫伯特·A. 西蒙 70

SIS (susceptible-infected-susceptible) model "易感-感染-易感"模型 102

six degrees of separation 六度分隔理论 5, 47—48, 54

small-world property 小世界属性 49—53, 62, 84

smelling gasoline epidemics (1998) 闻到汽油传染(1998) 106

snowball sampling 雪球式抽样 31

social capital 社会资本 31

social order 社会秩序 2

social spreading 社交传播 106—107

social structure 社会结构 10, 17

sociometry 社会计量学 10

software of life metaphor "生命的软件"隐喻 23

Sola Pool, I. de 伊锡尔·德·索拉·普尔 48

Solla Price, D. de 德里克·德·索拉·普莱斯 69

sparse networks 稀疏网络 45

stock price correlations 股票价格的相关性 36
Strogatz, S. *see* Watts-Strogatz model 史蒂文·斯托加茨，见 Watts-Strogatz model
structural balance 结构上平衡 21
structural equivalence 结构上等同 112
structural holes 结构洞 83
subgraphs 子图 92
suicide in adolescent girls 青春期女孩的自杀 83
superconnectors 超级连接器 55—65
super-spreaders 超级传播者 103
swine flu 猪流感 100—101
systemic failures 系统性故障 99

T

Tanzania laughter epidemics (1962) 坦桑尼亚的笑声传染病（1962）106
targeted immunization 针对性免疫 103
TCP/IP 传输控制协议／互联网协议 38, 39
terrorist attacks 恐怖袭击 12, 40, 55
three degrees rule 三度空间规则 86
top species 顶位物种 28, 56
topology 拓扑学 12, 15, 16—17, 78, 92, 108, 111—112, 114
transitive triples 可传递三元组 84
transitivity 传递性 84—85
transportation networks 交通网络 37, 87, 100—101
triangles 三角形 84

tricarboxylic acid cycle 三羧酸循环 74
trophic species 营养物种 14, 112

U

ultra-small worlds 超小世界 63
universality 普适 5
universality classes 普适类 113—114

V

vertices 顶点 9, 11, 13, 24, 33, 55, 63, 67, 70, 73, 78
Vespignani, A. 亚历山德罗·韦斯皮尼亚尼 100, 102
viral marketing 病毒式营销 50, 105—106
viruses 病毒 98—102
 computer 计算机 5—6, 49, 104—105
vulnerability 脆弱性 103

W

Wadlow, Robert 罗伯特·瓦德洛 57—58
Watts-Strogatz model 沃茨-斯托加茨模型 51—52, 53, 84—85
weak ties 弱连带 30, 53
weighted networks 加权网络 22
weights 权重 22
'winner takes all' "赢者通吃" 55
word association 词汇联想 33
words networks 语词网络 32—34
world trade web 世界贸易网 36, 53, 62,

68, 77—78, 87

World Wide Web (WWW) 万维网 5,
39—41, 43—44, 49, 73, 79, 96, 105, 111

Y

Youm, Y. 尤思科·尤姆 80, 82

Yule, G. U. 乔治·尤德里·尤尔 70

Z

Zachary, W. W. 韦恩·W. 扎卡里
88—90

Zuckerman, H. 哈丽雅特·朱克曼 68

Guido Caldarelli and Michele Catanzaro

NETWORKS

A Very Short Introduction

To my family
G. C.

To Anna
M. C.

Contents

   List of illustrations i

1 A network point of view on the world 1

2 A fruitful approach 7

3 A world of networks 23

4 Connected and close 42

5 Superconnectors 54

6 Emergence of networks 66

7 Digging deeper into networks 80

8 Perfect storms in networks 94

9 All the world's a net; or not? 110

   Further reading 115

List of illustrations

1 A partial food web for the 'Scotian Shelf' in the north-west Atlantic off eastern Canada **3**
Reprinted with permission from David Lavigne

2 An engraving of Königsberg (top) and a graph (bottom) **8**
(left) © 2012. Photo Scala, Florence/BPK, Bildagentur fuer Kunst, Kultur und Geschicte, Berlin
(right) © Universal Images Group limited/Alamy

3 A 'metric' representation of London Tube (top) versus a 'topological' one (bottom) **15**
© Transport for London. Collection of London Tranport Museum

4 Three different representations of the same graph **16**

5 A United Kingdom Grassland foodweb **29**
Courtesy of Neo Martínez

6 Bow-tie structure of a network with directed links **44**

7 Small-world network model **52**
Reproduced from D.J. Watts and S.H. Strogatz, Collective Dynamics of 'Small-World' Networks, Nature, 393, 440-442 (1998). Reprinted with permission of Macmillan Publishers Ltd.

8 A homogeneous network (left), compared to a heterogeneous one (right) **59**

9 Height is a homogeneous magnitude while the number of friends is a heterogeneous magnitude **60**

10 Preferential attachment mechanism for network growth **70**

11 Structure of friendships of a karate club studied by anthropologist Wayne Zachary **89**
Data from W. W. Zachary, An Information Flow Model for Conflict

and Fission in Small Groups, *Journal of Anthropological Research* 33, 452-473 (1977)

12 The 14th-century bubonic plague (left) verses 2009 swine flu (right) **101**

V. Colizza, A. Vespignani, The Flu Fighters, *Physics World*, 23 (2), 26-30, February 2010. Courtesy of the authors and *Physics World*

Chapter 1
A network point of view on the world

Networks are present in the everyday life of many people. On a typical day, we check emails, update social network profiles, make mobile phone calls, use public transportation, take planes, transfer money and goods, or start new personal and professional relations... In all these cases—consciously or not—we are using networks and their properties. Similarly, networks appear in important global phenomena. Financial crises generate domino effects in the web of connections between banks and companies. Pandemics—like avian flu, SARS, or swine flu—spread in the airport network. Climate change can alter the network of relations between species in ecosystems. Terrorism and war target the infrastructure grids of a country. Large-scale blackouts take place in power grids. Computer viruses diffuse in the Internet. Governments and companies can track people's identity through their social networks and other digital communication tools. Finally, the various applications of genetics depend on the knowledge of the genetic regulatory networks that operate within the cell.

In all these situations we deal with a large set of different elements (individuals, companies, airports, species, power stations, computers, genes...), connected through a disordered pattern of many different interactions: that is, they all have an underlying network structure. Often, this hidden network is the key to understanding those situations. A good example is the collapse of cod population in the north-western Atlantic in the eighties. At the time, the shortage of cod generated a massive economic crisis in the Canadian fishery industry. Canadian stakeholders asked for more expeditions to hunt seals, maintaining that controlling these predators of cod would help stop the collapse. Many seals were killed during the nineties, but the cod population did not recover. In the meantime, ecologists studied the different food chains that connected cod and seals. By the end of the decade, they drew a complete map (Figure 1) where a lot of different chains were found to connect the two species. In the light of this intricate picture, hunting for predators of cod does not necessarily help the fish. For example, seals predate about 150 species, and several of them are predators of cod: thus, reducing the population of seals can end up increasing the pressure of other predators on cod.

Ecosystems are complex webs of species: it is crucial to take into account this underlying network structure if we want to understand and manage them. Similar caution must be taken with several other systems, all based on a networked architecture. For example, the development of an infectious disease like AIDS is strongly influenced by the pattern of non-protected sexual relations within a population. Similarly, liquidity shocks depend on the interwoven network of money exchange between banks.

All the examples above are instances of the so-called *emergent phenomena*. That is some collective behaviour that cannot be predicted by looking at the single elements forming the system. Usually, systems that display these phenomena are dubbed *complex systems*. For example, a single ant is a relatively awkward animal, but many ants together are capable of activities as complex as building large anthills or storing large quantities of food. In human societies, social order arises from the combination of autonomous

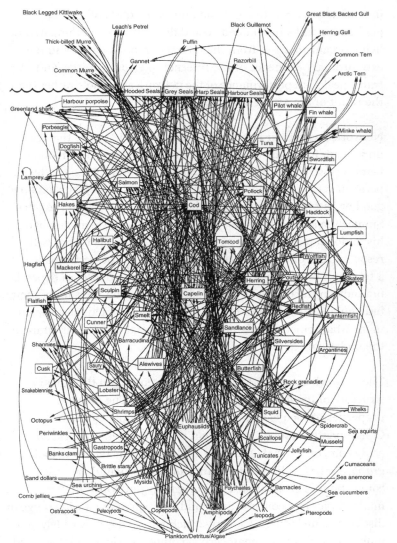

1. A partial food web for the 'Scotian Shelf' in the north-west Atlantic off eastern Canada. Arrows go from the predator species to the prey species

individuals, often with conflicting interests, that still end up performing tasks that nobody could do on their own. Similarly, a living organism arises from the interaction of its parts; and the extraordinary resilience of the Internet to errors, attacks, and traffic peaks is a performance of the network as a whole rather than the result of the action of individual machines.

Networks, with their emphasis on the interactions, are the key to understanding many of these phenomena. Imagine two football teams whose players have very similar skills, and yet the two teams perform very differently: probably this difference depends on how good or bad the interactions are between the players on the pitch. Similarly, a single player can be good in his league team and bad in his national team because of the different positions he has with respect to the other players in the two groups. The performance of a team is a kind of emergent phenomenon, one that does depends not only on the quality of the single players or on the sum of their individual skills, but also on the network of interactions between them. Many emergent phenomena rely crucially on the structure of the underlying networks.

The network approach focuses all the attention on the global structure of the interactions within a system. The detailed properties of each element on its own are simply ignored. Consequently, systems as different as a computer network, an ecosystem, or a social group are all described by the same tool: a *graph*, that is, a bare architecture of nodes bound by connections. This approach was originally developed in mathematics by Leonhard Euler and later spread to a wide range of disciplines, including sociology, which has deployed it widely, and more recently physics, engineering, computer science, biology, and many others.

Representing widely different systems with the same tool can only be done by a high level of abstraction. What is lost in the specific description of the details is gained in the form of *universality*—that is, thinking about very different systems as if they were different realizations of the same theoretical structure. In this respect, the spread of a computer virus can be similar to flu; hacking a router can have the same effect as the extinction of a species in an ecosystem; and the growth of the World Wide Web (WWW) can be set alongside the increase in scientific literature.

This line of reasoning provides many insights. For instance, representing a system as a graph allows us to perceive large-scale structures that encompass apparently unrelated elements. In 2003, a trivial accident in the Swiss electric grid triggered a large-scale blackout affecting Sicily, 1,000 kilometres away. Focusing on the network structure allows us to see that faraway elements end up being strongly connected, through incredibly short paths of relation or communication. The current observation that two individuals geographically and socially apart—such as a rainforest inhabitant and a manager in the City of London—are connected by only 'six degrees of separation' is not far from reality, and it can be explained in terms of the network structure of social relations.

The network approach also sheds light on another important feature: the fact that certain systems that grow without external control are still capable of spontaneously developing an internal order. Cells or ecosystems are not 'designed' but nevertheless work in a robust way. Similarly, social groups and trends arise from an immense variety of different pressures and motivations but still display clear and definite shapes. The Internet and the WWW boomed without the presence of any regulating authority and were promoted by an enormous variety of unrelated agents: however, they usually work in a coherent and efficient way. All these are *self-organized processes*, i.e. phenomena in which order and organization are not the result of an external intervention or global blueprint but the outcome of local mechanisms or tendencies, iterated along thousands of interactions. Network models are able to describe in a clear and natural way how self-organization arises in many systems. As well, networks allow us to better understand dynamical processes such as the rapid spread of computer viruses, the large scale of pandemics, the sudden collapse of infrastructures, and the bursts of social phobias or music trends.

In the study of complex, emergent, and self-organized systems (the modern science of complexity), networks are becoming increasingly important as a universal mathematical framework, especially when massive amounts of data are involved. This is typically the case of individuals accumulating queries in search engines, updates in social websites, payments online, credit card data, financial transactions, GPS positions from mobile phones, etc. In all these situations networks are crucial instruments to sort out and organize these data, connecting individuals, products, news, etc. to each other. Similarly, molecular biology relies more and more on computational strategies to find order in the large amounts of data it produces. The same happens in many other fields of science, technology, health, environment, and society. In all of these, networks are becoming the paradigm to uncover the hidden architecture of complexity.

Chapter 2
A fruitful approach

Crossing Euler's bridges

In the Russian city of Kaliningrad, the island of Kneiphof stands in the Pregel River. Three centuries ago, the city was in Prussia, it was called Königsberg, and at that time seven bridges connected the island with the rest of the city (Figure 2 top). A riddle was popular in the town: was it possible to walk across the seven bridges, without crossing any of them twice? Nobody had ever been able to do this successfully. On the other hand, no formal proof was available about the feasibility of such a walk. The solution came from one of the most famous mathematicians of all times. In 1736, Leonhard Euler drew a map of Königsberg in an unusual fashion. He represented the portions of mainland and the island as dots, and the bridges as lines, connecting the dots with each other (Figure 2 bottom).

When we cast the problem in this form, things become easier. By showing the network out of the city, Euler proved that the walk was impossible. His explanation is based on the following observation: for such a walk to be feasible, all the dots along the path must have an even number of connections. This is because every time one enters any part of the city via a bridge, one must leave via a different bridge. In general, each area must have an even number of bridges, e.g. 2, 4, 6. Only the start points and end points of the walk can have an odd number of links: at the start point, there can be

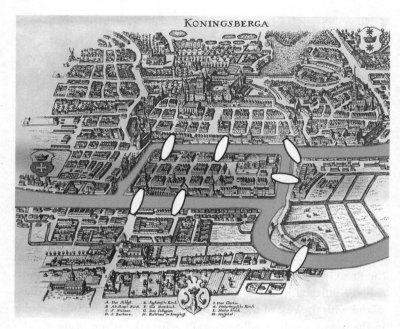

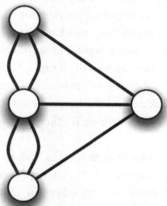

2. An engraving of Königsberg (top), now Kaliningrad, showing the Königsberg bridges riddle, represented by Leonard Euler as a graph (bottom)

only one bridge; and the same holds for the end point. Unfortunately, the graph of Königsberg has all the vertices with an odd number of links. Therefore it is impossible to walk only once over all the bridges.

Such a simpilified mathematical map of Königsberg is the first example of a *graph*. Mathematicians call the dots and lines forming it (respectively) *vertices* and *edges*. Nowadays, Euler is credited for having started a whole branch of mathematics, built on graph analysis. His intuition can be considered the first, foundational moment of network science. After him, many mathematicians studied the formal properties of networks, while scientists applied them to a wide range of problems: electrical circuits by G. R. Kirchhoff in 1845, isomers of organic components by A. Cayley in 1857, the mathematical concept known as the 'Hamiltonian cycle' by W. R. Hamilton in 1858, etc. One famous application is the 'map colouring problem', proposed in the mid-19th century. At that time, geographers were sorting out the minimum number of colours needed to draw maps, where adjacent countries have different colours. This was more than a theoretical problem: given the large number of countries and the reduced quantity of different inks available in the print industry, it was essential to be able to use the minimum number of colours. Empirically, three colours were not enough while four seemed to work well. The first proof that the solution was indeed four was obtained only in 1976. It is based on representing a map as a graph, where nodes are the countries and an edge is drawn between two of them if they share a border.

Runaway girls, Australian peoples, and Chicago workers

In the autumn of 1932, fourteen girls ran away in a period of just two weeks from Hudson School for Girls, in New York State. This was much more than the usual rate of runaways. School managers then decided to look into the individual personalities of the girls, to understand this phenomenon. Since no striking evidence of particular differences in the girls' personalities was found, the psychiatrist Jacob Moreno proposed a completely different explanation. He suggested that this large number was triggered by the position of runaways in the girls' social network. Moreno, together with Helen Jennings, mapped the social ties between the students, using *sociometry*, a technique for identifying relations between individuals. They found that these ties were the main channel through which the idea of running away spread among the girls. The position of an individual in the friendship network was crucial for replicating the behaviour of the runaway girls.

Moreno was one of the first researchers to apply the idea of network to society. After Euler's intuition, his work is a second and crucial step in network science foundations, starting one of the most important lines of network science: the analysis of social networks. Thirty years later, anthropologists applied a similar approach to kinship relations in tribes such as the Arunda people in Australia. In this case, the connections were drawn between individuals who were relatives. Researchers found that the resulting diagrams responded to elegant mathematical structures. These and other results suggested that neat social structures, or even universal laws, could be found underneath the disorder of the human social world. Since then, social sciences have widely deployed the concept of network to represent social structure. Many other studies followed these seminal ones, focusing on the circles of women in southern America (by A. Davis, B. B. Gardner, and M. R. Gardner in 1941), groups of factory workers in Chicago (by E. Mayo in 1939), friendship between schoolchildren (by A. Rapoport in 1961), and the relations between drug addicts in the city of Colorado Springs (by Richard B. Rothenberg et al. in 1995), among others.

Random connections

The third important moment in the foundation of network science came with the publication of a set of papers by the two mathematicians Paul Erdős and Alfréd Rényi, in 1959–61. Erdős, one of the most important mathematicians of the 20th century, has been described as 'a man who loved only numbers'. Partly wrong: he loved graphs too. The two theorists studied a mathematical model representing a graph where vertices are connected to each other completely at random. This model, lately known as Random Graph, was first proposed in 1951, in a paper by mathematics researchers Ray Salomonoff and Anatol Rapoport.

The random graph is a very simplified model, and its properties are very different from those of real networks. For example, randomness and chance could indeed play an important role in meeting new friends, but the formation of friendship networks is certainly related to many other factors, such as social class, common languages, affinity etc. However, the random graph model is very important because it quantifies the properties of a totally random network. Random graphs can be used as a benchmark, or *null case*, for any real network. This means that a random graph can be used in comparison to a real-world network, to understand how much chance has shaped the latter, and to what extent other criteria have played a role.

The simplest recipe for building a random graph is the following. We take all the possible pairs of vertices. For each pair, we toss a coin: if the result is heads, we draw a link; otherwise we pass to the next pair, until all the pairs are finished (this means drawing the link with a probability $p = \frac{1}{2}$, but we may use whatever value of p). Typically, the creation of the graph and its study are not done manually. Scientists use computer programs and draw the resulting network on paper or on computer screen. However, this becomes increasing difficult when the network is large. Moreover, it is hard to study an interwoven structure just by visual inspection. A better and quantitative insight comes from studying the graph as an abstract object, by means of mathematical tools. Computers can also help: simulation allows us to build, within the 'mind' of a computer, a faithful realization of the model, and then make measurements of it as if it were a real object. If we want to compare the abstract model with a real-world network, now we just need to compare the measures in both cases.

Since its introduction in the sixties, the random graph model has become one of the most successful mathematical models, despite its loose connection with reality. Nowadays it is a benchmark of comparison for all networks, since any deviations from this model suggest the presence of some kind of structure, order, regularity, and non-randomness in many real-world networks.

Nets to fish for information

In the wake of the terrorist attacks in New York in 2001, in Madrid in 2004, and in London in 2005, several governments have proposed, as an anti-terrorist measure, the storage of electronic traffic data. In this proposal, years of phone calls and emails between citizens would be recorded for safety purposes. The content of the messages would not be stored: it would be enough to record only senders and recipients (and sometimes time and place of communication). As police forces know, even this simple map of who is connected with whom is a powerful instrument in tracking people's activity. Indeed from a sketch of phone calls, we can deduce habits, circles of friends, and various other data about a person.

This is a very practical example—and a very debated one—of the basic approach of network science. A complex system is represented as a graph—a set of equal elements connected by equal interactions—ignoring the detailed features of its constituents, and the specific nature of their relations. This procedure may seem too drastic, but still it allows us to capture more information than expected at first sight. A proof of the effectiveness of this approach is the friendship advice system contained in many online social networks, such as Facebook or LinkedIn. The idea is simple: you are likely to know the friends of your friends. As simple as it is, it works most of the times. This method, called *educated guess*, is also behind the systems that suggest books or other items in online shops. The software of these companies exploits networks of goods associated with each consumer. This is why commercial companies store a great deal of electronic data, including email messages and online social network data: they know that this information is extremely useful.

The basic approach of network science can be applied to a broad set of systems. For example, an ecosystem composed of hundreds of species, where each one extracts energy from others by means of different strategies of predation. In the network approach, this is represented by a set of identical vertices connected by a set of identical edges. The same strategy is applied to systems ranging from the Internet (hundreds of thousands of computers, routers, exchange points, etc., connected by telephone lines, fibre optic cables, satellite communications, etc.) to a human population (a huge number of actors with different purposes and roles, connected by a variety of relations). While the network approach eliminates many of the individual features of the phenomenon considered, it still maintains some of its specific features. Namely, it does not alter the size of the system—i.e. the number of its elements—or the pattern of interaction—i.e. the specific set of connections between elements. Such a simplified model is nevertheless enough to capture the properties of the system.

From individuals to groups

There are two possible approaches when dealing with a situation where many different elements interact in different ways. In the first approach, we identify the basic constituents and interactions between them. By studying each element on its own, we can then deduce the behaviour of the system as the sum of its individual elements. For example, ecologists can describe the features of an ecosystem by listing prey and predators of every single species. Computer scientists describe a network of computers by focusing on the features and protocols of each different machine. Psychologists study social relations by describing the behaviour of each social actor in his or her circle.

A second strategy, different from the first one, consists in putting many elements together into a few homogeneous groups. For example, sociologists and political scientists usually split society into social classes, genders, levels of education, ethnicities, nations, etc. Similarly, epidemiologists often separate the population into a limited set of 'compartments': healthy individuals, infected, immunized, etc. Ecologists can also use this approach by aggregating into groups (*trophic species*) all the species that have similar roles in a foodweb.

The network approach tries to complement these two points of view. Many phenomena are impossible to explain if one focuses only on the behaviour of individual elements. For example, the dynamics of a species population within an ecosystem do not depend on the features of that species alone: the global network of prey–predator interactions must be taken into account. On the other hand, focusing on big classes of elements may not be useful either. For example, the political phenomena taking place in a country are hardly the outcome of a pre-existing national identity but rather of the specific pattern of social relations within that country. The network approach is somewhere between the description by individual elements and the description by big groups, bridging the two of them. In a certain sense, networks try to explain how a set of isolated elements are transformed, through a pattern of interactions, into groups and communities. In all cases where this pattern is relevant, the network approach provides essential insights.

Geography versus 'netography'

At the beginning of the 20th century, London's underground train service (the Tube) became so intricate that more and more complicated maps had to be issued from time to time, in order to orient the travellers. In 1931, after many attempts, Henry Beck, an employee of the company, changed the criteria for drawing the chart. Instead of embedding the lines on top of an actual map of London, Beck placed them in an abstract space (Figure 3).

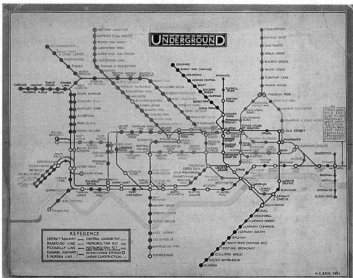

3. A 'metric' representation of the London Tube (top) versus a 'topological' one (bottom). Even though in the latter the actual positions and relative distances of the stations are not indicated, it is a more useful mind map of the service

Stations were represented by well-spaced dots. Tube connections became straight lines with neat angles of 45 or 90 degrees. This map has little to do with the real positions and distances of stations, but it is much clearer and more useful for the passengers. Those travelling on the Tube network are not interested in its geographic features: the information about the sequence of stations and the intersection of Tube lines is enough.

Henry Beck's London's Tube map is basically a graph. His solution to the mapping problem exploited a basic feature of the network approach: in networks, *topology* is more important than *metrics*. That is, what is connected to what is more important than how far apart two things are: in the other words, the physical geography is less important than the 'netography' of the graph. The difference between these two concepts is shown in Figure 4. The three images represented in the picture are different from a metric point of view. That is, the positions of nodes in space and the lengths of links are different. However, from the topological point of view they are identical: they are just three different representations of the same graph. In the network representation, the connections between the elements of a system are much more important than their specific positions in space and their relative distances.

The focus on topology is one of its biggest strengths of the network approach, useful whenever topology is more relevant than metrics. For example, an email sent from New York reaches an office in London in the same time as one sent from the office next door. Even on the Internet, a material infrastructure embedded in geographical space, the pattern of the connections is more

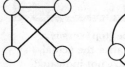 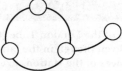 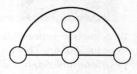

4. Three different representations of the same graph

important than the physical distance. In social networks, the relevance of topology means that *social structure* matters. Lionel Messi is nowadays one of the best football players in the world. However, his performance differs according to which team he is playing in (either Argentina or FC Barcelona). Some social scientists have argued that the network of Messi's relations with other players in Argentina is different from that in FC Barcelona. According to their research, this results in the player carrying a heavier 'weight' in the former case: this may explain, at least in part, the difference in his performance. Similar phenomena appear also in more complicated social 'games', where an individual's outcome can be strongly influenced by his or her position in a network of relations.

Chains, grids, and networks

The network approach reduces complex systems to a bare architecture of nodes and links. This is a substantial simplification, but still the resulting graph may not be so easy to interpret: this is the case with the tricky illustration shown in Figure 4. Even a graph as simple as an innocent chain of nodes can be a rather complicated object to handle. A chain may represent, for example, a fire brigade moving a bucket of water; or a food chain of species, in which the first predates the second, which predates the third, and so on; or a business-to-business supply structure: a set of companies in which each one supplies the next one.

Imagine a production chain of five companies (1, 2, 3, 4, and 5). Along this chain, any of them can make a deal with either of its two neighbours. The rule is that each company can close only one contract: for example, if 3 closes a deal with 2, it cannot have arrangements with 4. Given this simple structure and rule, it turns out that nodes 1 and 5 have less bargaining power, since they have fewer alternatives. This makes nodes 2 and 4 stronger, and (unexpectedly) it weakens node 3. Indeed, node 3 has only strong nodes to deal with, and therefore it ends up having less convenient deals. Something as simple as a linear sequence of nodes does indeed yield a rather complex landscape. This example shows what sociologists call an *exclusion mechanism*. Far from being a theoretical situation, this is commonly experienced in economics, when the establishment of a commercial relation between two parts excludes a third node.

To further complicate the question, one has to take into account that real-world systems are rarely as simple as a chain. In the past, scientists have represented complex systems through regular grids or *lattices*, instead of using graphs. These objects are composed of many elements—representing people, animals, computers, etc.—usually arranged along a regular pattern of connections, like pieces on a chessboard connecting only with their four neighbours. This regular structure makes systems much easier to handle by mathematical calculation and by computer simulation than when using a graph.

The lattice choice, albeit simpler with respect to a graph, introduces nevertheless a strong limitation. In fact, a lattice is suitable only for representing systems carefully designed or subject to strong constraints. These might be, for example, the arrays of processors in a computing cluster or verbal communications between chain workers in a noisy workplace. In lattices, every node is linked to a fixed number of nearest neighbours, while in the vast majority of real-world cases, links connect a variable number of elements, no matter whether close to or far away from each other. The ability to capture this disorder is one of the great advantages of the network approach.

Much of this disorder is encoded in a crucial quantity: the *degree*, that is, the number of edges attached to each node. If a node is a web page, the degree represents the number of links it receives from other pages. If a node is a species, the degree is the number of species it depends on for feeding. If a node is an individual, the degree is the number of acquaintances. This circle can be related to what sociologist Peter Marsden has called a *core discussion network*: the set of people (friends, partners, family members, current and past schoolmates, co-workers, neighbours, fellow members of clubs, professional advisers, consultants, etc.) with whom one discusses important matters or spends time.

Mapping relationships

Two people can have an infinite set of possible relations. They may share attitudes, ideas, or gender. They may be friends, relatives, or co-workers. They may be sexual partners or simply play in the same football team. Furthermore, two or more of these connections may simultaneously occur between the same couple of people. Some of them are cooperative relations, while others convey open hostility, with an entire spectrum possible in between. Finally, some can be perceived only by one side and ignored by the other (e.g. the fans of a rock star feel linked to him, while the star may simply ignore them). Sociology has classified a broad range of possible links between individuals (Table 1). The tendency to have several kinds of relationships in social networks is called *multiplexity*. But this phenomenon appears in many other networks: for example, two species can be connected by different strategies of predation, two computers by different cables or wireless connections, etc.

We can modify a basic graph to take into account this multiplexity, e.g. by attatching specific tags to edges. For instance, we can take into account whether a connection is positive or negative. Species are linked by predation (negative), but also by mutualism (e.g. the positive relation established between flowering plants and pollinators). People can be enemies (negative) or friends (positive). A web page can link to another web page to criticize its content (negative) or to advertise it (positive). Adding this simple binary feature complicates things a lot. Imagine a group of three people, Alice, Bob, and Carol. When positive

Table 1. A classification of ties in social networks.

Similarities			Social relations					Interactions	Flows
Location	Membership	Attribute	Kinship	Other role	Affective	Cognitive			
e.g.,	e.g.,	e.g.,	e.g.,	e.g.,	e.g.,	e.g.,		e.g.,	e.g.,
Same spatial and temporal space	Same clubs	Same gender	Mother of	Friend of	Likes	Knows		Sex with	Information
	Same events	Same attitude	Sibling of	Boss of	Hates	Knows about		Talked to	Beliefs
				Student of		Sees as happy		Advice to	Personnel
				Competitor of				Helped	Resources
								Harmed	

relations connect all of them, everything is fine. Alternatively, Alice and Bob may be linked by friendship but both have hostile relations towards Carol. Things get complicated when the situation is reversed: Alice has positive relations with both Bob and Carol, but the two of them hate each other. And things get really problematic when everybody hates everybody else. According to sociology, the first and the second situation are *structurally balanced*, while the third and the fourth are not. In 2006, mathematician Tibor Antal and physicists Paul Krapivsky and Sidney Redner applied this concept to the shifting diplomatic alliances between six European countries before the First World War. They showed that their alliance gradually evolved into a structurally balanced situation, where either strong alliances were established or clear common enemies were identified. The six countries became divided into two groups (on one side, Britain, France, and Russia; on the other, Austria-Hungary, Germany, and Italy), each of them allied to all the countries in its group and enemy to all the countries in the other group. Soon after this situation had come about, the war broke out. This example shows that structural balance is not necessarily something desirable.

Graph theory allows us to encode in edges more complicated relationships, as when connections are not reciprocal. Wolves predate sheep, blogs link to large newspapers, and some people fell in love with others; the reverse is seldom true. In this case, the connections in the graph are a sort of one-way street where we can travel in one direction but not backwards. If a direction is attached to the edges, the resulting structure is a *directed graph* where links are indicated by an arrow. In these networks we have both *in-degree* and *out-degree*, measuring the number of inbound and outbound links of a node, respectively.

The relations considered so far are binary: that is, they acquire only two values. Such *dichotomous* connections either exist or do not: for example, being married to somebody or being employed by somebody. Statistically, however, these are exceptions: in most cases, relations display a broad variation of intensity. Predation is counted in the number of prey eaten, web pages can be connected by a sporadic link or by a large number of connections, and love can range from slight attraction to furious passion. Such further features correspond to the *weights* we can add to the links. *Weighted networks* may arise, for example, as a result of different frequencies of interaction between individuals or entities.

Other modifications of the basic graph structure are possible, and the techniques to handle these objects are very interesting. For example, a large part of social network research is devoted to working out how different kinds of ties affect each other. However, the strength of the network approach is that, in some cases, it is justifiable and effective to ignore all or most of the specific details: directed networks become undirected, weights are removed, multiple links are collapsed in a single edge, etc. Results show that this radical simplification can still capture a remarkable amount of information.

Chapter 3
A world of networks

Networkomics

During the eighties and nineties of the last century everything was 'genetic' in some way. Newspapers published stories about 'the gene for homosexuality', 'the gene for obesity', 'the gene for violence', or the 'gene for alcoholism'. This attitude responded to the expectation that the secret of human complexity was hidden in the genome. The DNA—the deoxyribonucleic acid molecule packed in the nucleus of the cell, that contains the genes—was dubbed the 'software of life', the program responsible for every single feature of a living being, the code whose dysfunction caused all the diseases. This vision set off a rush to sequence the genome, culminating in the publication of its map in February 2001. Results were quite surprising. Human beings do not have many more genes than a nematode worm, and fewer than some species of rice. It is reasonable that the human genome is almost identical to that of great apes, but the problem is that it is also rather similar to that of mice. The software metaphor did not stand in the face of this evidence: the DNA sequence alone does not explain the observed differences between species, let alone all the features and diseases of a single individual. In fact, there is a long series of steps from the genes to the macroscopic features of a living being. Variations in this path determine different outcomes.

The first layer of complexity above the gene level is given by *gene regulation*. Genes contained in the DNA are transcribed and translated to produce proteins. Proteins play a central role in almost every aspect of life: muscle movement, blood circulation, acting as enzymes, binding to hormones, etc. Moreover, proteins interact with each other: the production of a protein can be facilitated or hindered by the presence of other proteins in the cell. The delicate balance of these reciprocal influences is crucial for life. For example, the mutations of a single protein, the p53, are implied in a large number of different cancers. These interwoven patterns of activation and inhibition yield the *gene regulatory network*. In this net, nodes are genes and links are chains of reactions that connect the expression of a gene with that of others.

Protein interact, in all the forms in which they can happen, represent a second layer of complexity. For example, several proteins can bind together. These macromolecules behave as molecular machines, performing functions in the machinery of the cell. To do so, they must have the correct geometrical shape to fit with each other. When a protein is folded in the wrong way, several problems can arise. For example, the proteins responsible for the 'mad cow syndrome' (Creutzfeldt-Jacob disease) in humans, i.e. the *prions*, are supposed to be nothing else than misfolded proteins. All the possible physical connections between proteins can be represented as a network. In the *protein interaction network* the vertices are proteins and an edge is drawn between them if they physically interact in the cell.

Proteins are not enough for making a cell work. The cell interchanges matter, energy, and information with the environment, through many different molecules, involved in millions of reactions. Hunger, satiety, coldness, and in general all the states experienced by the organism, depend on this set of reactions, called *metabolism*. The chains of reactions that convert one molecule into another, passing through a series of intermediates steps, are called *metabolic pathways*. However, reactions in cells rarely follow the pattern of an ordered sequence. For example, the final molecule often interacts with the initial one in order to stop the reaction. This feedback process closes a loop in the chain of reactions. The ensemble of all such paths yields an intricate *metabolic network*.

An organism is therefore the outcome of several layered networks and not only the deterministic result of the simple sequence of genes. Genomics has been joined by *epigenomics, transcriptomics, proteomics, metabolomics*, etc., the disciplines that study these layers, in what is commonly called the *omics revolution*. Networks are at the heart of this revolution.

Thinking webs

The idea that the 'soul' could be embodied in an organ sounded a weird supposition by the 18th century. But physicians were aware that a stroke or other brain injuries could compromise crucial cognitive functions: the link between mind and brain was then starting to become evident. At that time, the anatomist Franz Joseph Gall dared to propose that all mental functions must arise from the brain. He identified 27 'organs' within the brain, each one responsible for colour, sound, memory, speech, as well as friendship, benevolence, pride, etc. The idea sounded so heretical that Gall had to flee Vienna and find shelter in revolutionary France.

Later on, several physiologists tried to verify Gall's theory, for example by removing slices from pigeons' brains. However, they could not find any evidence of the organs that Gall postulated. For this reason, they arrived at the conclusion that the brain was a homogeneous, undifferentiated unity that generated thought: 'the brain secretes thought as the liver secretes bile', as one of them put it. This conception dominated until the studies of Paul Broca in the 1860s. In autopsies of patients with expressive aphasia, Broca always found some damage to the frontal lobes of the left side of the brain.

'We speak with the left hemisphere,' he declared, after having identified what is now called *Broca's area*. Since then, neurologists have found various centres responsible for different activities, but they have also found that they rarely work in isolation: the integration of different areas of the brain is crucial to its functioning.

Networks provide a bridge between the paradigms of a brain divided into specialized areas versus the model of the brain as a whole (not dissimilar to what happens in social sciences, where networks allow us to describe society at a level between individuals and communities). The brain is full of networks where various web-like structures provide the integration between specialized areas. In the cerebellum, neurons form modules that are repeated again and again: the interaction between modules is restricted to neighbours, similarly to what happens in a lattice. In other areas of the brain, we find random connections, with a more or less equal probability of connecting local, intermediate, or distant neurons. Finally, the neocortex—the region involved in many of the higher functions of mammals—combines local structures with more random, long-range connections. Some scientists think that these wiring schemes may be responsible for subjective awareness: the emerging conscience may be the result of a sufficiently complex network structure.

Pinning down the actual structure of these neuronal networks is extremely hard, due to the enormous quantity of cells and the difficulty of probing them. Only for very simple organisms such as the nematode worm, *Caenorabditis elegans*, we have a detailed map. This one-millimetre long, transparent creature, with a three-week lifespan, has only about 300 neurons but it is a superstar in molecular biology. *C. elegans* is a *model organism*, an animal that is especially suitable for experiments, because scientists know its features well, and some aspects of it are comparable to those of the human organism. This translucent worm is often the first benchmark for the trial of new medicines and treatments.

Drawing a similar neuronal network for the human brain is impossible at the moment. However, another strategy can be applied. When humans perform an action, even one as simple as blinking, a storm of electrical signals from the neurons breaks out in several areas of the brain. These regions can be identified through techniques such as *functional magnetic resonance*. Through this technique, scientists have discovered that different areas emit correlated signals. That is, they show a special synchronization that suggests that they may influence each other. These areas can be taken as nodes and an edge is drawn between two of them if there is a sufficient level of correlation. Also at this level, the brain appears as a set of connected elements. Each action of a person lights up a network of connected areas in their brain.

The blood vessels of Gaia

In 1999, San Francisco Bay experienced massive algal blooms. Normally, such blooms are the result of an intensive agricultural use of land: when we drain fertilizers such as nitrogen and phosphorus into the sea, they become nutrients for algae. However, this was not the case in this instance, since a number of policies and controls had decreased the level of nutrient pollution entering the bay from its various rivers. Compiling data from three decades of observations, ecologists in California concluded that the blooms had a much more complicated explanation. In 1997 and 1998, one of the strongest *El Niño* events was recorded, followed by an equally strong *La Niña* in 1999. These phenomena induced changes in the California current system. Deep, cold, and nutrient-laden waters emerged along the coast. Such nutrients attracted ocean dwellers—flatfish and crustacean—into the bay. These animals are predators of bivalves of the bay that, in turn, act as an obstacle to the spreading of algae. The collapse in the bivalve population due to the increase in its predator was the immediate cause of the algal blooms. The conditions that triggered this domino effect may be due to normal climatic fluctuations. Nevertheless, its consequences are a warning: climate change, and especially the increased frequency of extreme events, can have rather unexpected effects on ecosystems.

The central structure behind the San Francisco Bay algal bloom is a *food chain*—that is, a series of species in connection: flatfish and crustacean prey on bivalves, and bivalves consume algae. Through food chains, living organisms extract from each other the energy and matter they need to survive (this is not the only possible interaction between species in an ecosystem: organisms can also establish mutually beneficial interactions, such as those between flowering plants and their insect pollinators). Every food chain starts with *basal* species, such as plants and bacteria. These do not prey on any other species and take resources directly from the environment by transforming light, minerals, and water. These resources are transferred along the food chain by successive predations. *Intermediate* species are organisms that are both predators and prey. And *top* species (at the end of the chain) are those that are not predated by anything. Food chains help us understand why fisheries collapse, as the Peruvian anchovy fishery in the seventies. After periods of massive indiscriminate fishing, the result is a dramatic reduction of predators such as cod or tuna. After this phase, fishing tends to move towards more basal species, such as anchovies. But these rapidly collapse as well. The reason is that, when large predators are removed, they are replaced by other predators downstream in the foodweb. These are often non-edible fish: without population control, they deplete the other edible basal species.

The actual picture of an ecosystem is even more complicated: typically, food chains are not isolated, but interwoven in intricate patterns, where a species belongs to several chains at the same time. For example, a specialized species may predate on only one prey (or in some cases on only a few). If the prey becomes extinct, the population of the specialized species collapses, giving rise to a set of *co-extinctions*. An even more complicated case is where an omnivore species predates a certain herbivore, and both eat a certain plant. A decrease in the omnivore's population does not imply that the plant thrives, because the herbivore would benefit from the decrease and consume even more plants.

As more species are taken into account, the population dynamics can become more and more complicated. This is why a more appropriate description than 'foodchains' for ecosystems is the term *foodwebs* (Figure 5). These are networks in which nodes are species and links represent relations of predation. Links are usually directed (big fishes eat smaller ones, not the other way round). These networks provide the interchange of food, energy, and matter between species, and thus constitute the circulatory system of the biosphere. They are the blood vessels of the Earth.

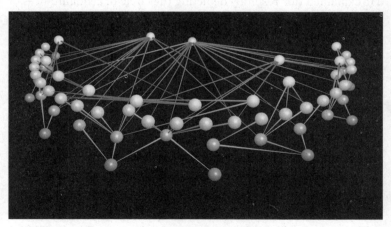

5. **A United Kingdom Grassland foodweb: nodes represent species of grassland plots in England and Wales, and links are drawn from predators (thicker end) to preys (thinner end)**

Homo 'retiarius' (Net man)

Word of mouth is a common way to obtain information about white-collar job openings. So if we are looking for this kind of job, it is a good idea to spread the word between friends and relatives. Less obviously, it may be even better to inform distant acquaintances and people we do not see often. This is what Mark Granovetter suggested in 1973. This sociologist interviewed a sample of professionals in a Boston suburb who had recently relied on personal contacts to obtain their jobs. He asked them how often they saw the person before obtaining the job. The majority reported 'occasionally' and a significant fraction answered 'rarely'. Job offers are more likely to come from old college friends, past workmates, and previous employers, than from close friends. Chance or mutual friends were the channels by which these connections were rediscovered. Granovetter described this phenomenon as *the strength of weak ties*.

He explained this result by depicting the circle of acquaintances of a hypothetical individual called Ego. Ego lives every day with his family and some close friends. Probably all these persons are also in close contact with each other. As a result, information travels fast in the group. So Ego is likely aware of all the news available in the group. On the contrary, weak ties connect him to faraway people. These individuals are not bounded by Ego's social surroundings. Therefore they open a whole set of new groups to him, each of them encapsulating information otherwise inaccessible. Missing the opportunity of weak ties causes difficulties in organizations, companies, or institutions. Information and skills become trapped in one group, without reaching those who need them. So these things have to be reinvented or paid for from outside consultants. A former CEO of HP is reported to have lamented: 'If HP only knew what HP knows!' Granovetter's intuition was later developed into the theory of *social capital*. This idea implies that the contacts of one person (and the contacts of these contacts) enable him or her to access resources that ultimately provide such things as better jobs and faster promotions. More generally, the position of an individual in his or her social network is crucial to determine future opportunities, constraints, outcomes, etc.

Measuring acquaintance is not easy, since it is such a subjective issue. Usually, maps such as the flowchart of a company are not very useful, because they do not correspond to the actual relations between workers. For this reason they are useless in helping us understand the channels (and possible bottlenecks) of information within the company. Scientists have devised a large set of alternative strategies to draw social networks, ranging from questionnaires to *snowball sampling*, a system in which an interviewed subject suggests somebody in their circle for the next interview. These strategies have enabled the collection of data as sensitive as the map of sexual intercourse between groups of different individuals (from high-school students in the US Midwest to people living in a village of Burkina Faso): the knowledge of these networks allows for better understanding of the spread of sexually transmitted diseases.

Another kind of relation that is relatively easy to pin down is professional collaboration. Such networks exist in several fields, ranging from Hollywood (two actors become connected if they play in the same movie) to science (two scientists become linked if they write a paper together). Collaborations can be found in more exotic environments such as politics (US senators have been connected on the basis of the co-sponsorship of laws) or terrorism (activists are connected on the basis of intelligence reports and legal documents).

Information technology provides a new and powerful way to measure interaction between people. Frequent phone calls and emails between two individuals, or friendship in virtual social networks like Facebook or LinkedIn, indicate a stable relationship and therefore an edge. More and more companies exploit the social networks of their customers to find such information. For example, telephone companies are reported to target 'influential' individuals with offers and other strategies: these are the customers who, when they change company, trigger similar changes in their close connections.

Webs of words, webs of ideas

'What shall King Henry be a pupil still / Under the surly Gloucester's governance?' says Queen Margaret in Part II of Shakespeare's King Henry VI trilogy (Act I, Scene III). The queen complains about the influence of the duke of Gloucester on her husband, the king. What does she mean by 'pupil'? A thesaurus suggests that *pupil* can be rephrased as scholar, acolyte, adherent, convert, disciple, epigone, liege man, partisan, votarist, or votary. This list provides a full spectrum of words to denote one person subjected to another. We can enlarge the spectrum by exploring all the words related to 'pupil' that is, its *semantic area*. These include faithful, loyalist, advocate, backer, supporter, satellite, yes-man... What role does the queen desire for her husband? Antonyms of 'pupil' suggest non-student, coryphaeus, leader, apostate, defector, renegade, traitor, and turncoat. Naturally, she is asking King Henry to rebel against the authority of the duke of Gloucester.

This is a simple example of how words link to each other. Indeed, a rigorous analysis should take into account the historical differences in the use of a word, the specific occurrence of a term in Shakespeare's work, the context in which it is used in the script, and many other aspects. In any case, it is fair to say that we can better understand the meaning of a word if we take into account its 'neighbours' in language. *Synonimity, antonymity*, and *semantic connection* are just a few of the possible relations. Others are *meronymity* and *hypernimity* ('beer' is a meronym of 'drink' and the latter is a hypernym of the former). 'Pupil' provides a crystal-clear example of another kind of connection. This term has two completely different meanings: it indicates both a student and a part of the eye. It is a case of *polysemy*. Of course, the context of Shakespeare's dramas immediately determines the correct meaning. In general, context provides the exact meaning of words: the *co-occurrence* of words in sentences defines their meaning. Such a co-occurrence provides yet another relation between words. For example, the words 'king' and 'Henry' are much more likely to appear together in English sentences than 'king' and relativity, for example.

We can now create our own maps of language. We use words as vertices and the edges connect synonyms, antonyms, and polysemic words (these relations can be drawn from a thesaurus or a dictionary, while patterns of co-occurrence can be drawn from large language databases, such as the *British National Corpus*). Semantic connections are more difficult to pin down: their study forms a complete area of linguistics. Some languages have special dictionaries that associate one word with a set of related ones. An alternative approach is experimental *word association*. A word is provided to a sample of people, asking them to say the first word that comes into their mind after hearing it. The resulting words are then used to repeat the association experiment. Proceeding in this way, step by step we build our web of associations. Different instances of word networks display different results. These depend on the language, on the kind of text, on the education of the author of the text, or may be related to linguistic dysfunctions.

Word networks contain a lot of information, but usually they are not particularly useful in studying the actual content of the texts and the relations between the ideas expressed in different texts. This is a crucial issue for web queries, for example. Usually, complicated algorithms have to be implemented to perform this task. However, in some bodies of texts, a very precise network can be drawn. This is the case with scientific literature. Producing knowledge is never a solitary endeavour. A scientist is always 'a dwarf standing on the shoulders of a giant', as philosopher Bernard of Chartres is said to have pointed out for the first time in the 12th century. Scientists' work almost always builds on previous results. Researchers recognize this by citing several older publications at the end of their papers. Citations provide recognition of relevant results of the past, give credibility to new results and make reference to facts, technologies, and experiments that are accepted as valid, or criticized, within a work. Publications have been extensively standardized in recent years: articles are mostly in English, control methods have been homogenized (mainly through *peer review*), measures of impact have been devised, etc. At the same time, large electronic databases of publications have been established, with thousands of new items being added every day: articles, books, patents, projects, etc. All this produces a large network of publications: two items are connected if one of them cites the other. We can also identify authorship from these databases and create networks of collaborations between scientists. These systems are increasingly used to map and visualize the development of knowledge and the most active areas of science.

Money by wire

In 2008 a number of large financial institutions in the United States suddenly went bankrupt. In a few months, the majority of the developed world was involved in one of the largest financial crises ever seen. Much has been written on the causes of this crisis. What is certain is that it showed that economies are very tightly interconnected at a global level.

Classical economic theories represented economic actors as independent, completely rational agents, focused on maximizing their income. However, facts show that individuals, companies, institutions, and countries are not independent: everyone influences each other in many ways. Their behaviour, far from being completely rational, is strongly dependent on subjectivity, emotions, and reciprocal influence.

Lending money is one way in which companies and institutions can become tightly interconnected. An interesting case is the money exchanged on a daily basis between private banks so that it can be available to meet the possible requests of bank clients (becoming therefore *more liquid*). If customer requests should exceed the liquidity reserve of a bank, that bank can ask other banks to lend money. Central banks worldwide require other banks to place a part of their deposits and debts with them, to create a buffer against liquidity shortages. In this sense, central banks ensure the stability of the banking system, thus avoiding liquidity shocks. The freezing of the interbank lending network was one of the first signals of the 2008 financial crisis.

An even stronger relation than lending money is given by *shareholding*, i.e. the direct participation of a company in another company's capital. This means that the first company holds a part of the second one, and can exert influence on its main decisions. Shareholding is converted into control when a company holds the majority of stock, or when it is able to determine the vote of the majority of the board. In this case, legally independent companies are converted into a business group. Often, these groups display a pyramidal structure, where a *holding* company is at the top, and operating companies are at the bottom of the control hierarchy.

The existence of business groups is explicit and legally regulated in most countries; but softer and less regulated forms of influence can exist. The most common of these happens within boards. Managers often sit on many boards at the same time. Obviously, they act as channels of information, alliances, or interests between boards. Their simultaneous presence in different boards establishes an *interlock* between their companies. If the companies are explicit competitors, this situation is clearly incompatible with a free market. A shared director will either favour one of the companies against the other or establish a cartel between them (which is generally ruled out by law). In general, such a director will find it very difficult to operate in the interest of all the investors in the different companies.

Further evidence of interconnectedness between companies is given by *stock price correlations*. Finance practitioners know that the stocks of companies operating in the same sector (e.g. mining, transport, services, food, etc.) change their prices in a somehow similar or 'synchronized' fashion. For example, stock prices of all the companies within the electronic sector (or any other) tend to decrease or increase at the same time. Financial analysts are interested in knowing how much of the change in a stock price is influenced by the change of another stock (in short, they want to know the *correlation* between stock prices). If this connection is strong enough, it is likely that the two companies are somehow connected. Lending money, shareholdings, shared directors, or stock price correlations are the main criteria by which a network between companies can be built: edges are drawn when one of these situations occurs.

The interconnectedness goes far beyond companies within one specific market. As the financial crisis has shown, events rapidly spread from national markets into the global scenario. One obvious channel by which this can happen is the import/export trade relationships among countries. This *world trade web* is a network in which nodes are countries and their trade relations define the edges. Like the cell, economies depend on these multiple layers of networks.

Critical infrastructures

On the night of 28 September 2003, lights went out across the whole of Italy, with the single exception of the island of Sardinia. It took several hours, in some places even days, to reinstate normal supply. Investigation showed that the blackout was triggered by a tree flashover close to a high-tension line between Italy and Switzerland. The resulting shortage of electric supply caused a sharp increase in demand on the remaining lines. As a result those lines collapsed, generating a ripple effect through the entire system.

Large-scale power outages reveal the connectedness of power grids. These systems deliver electricity across large distances from central points to cities and industrial areas. Carefully planned in the beginning, they grow more and more intricate over time. Nowadays, generators, transformers, and substations, connected by high-voltage transmission lines, constitute a network that spans several regions and often several countries (as the 2003 example shows). It is clear that such networks require careful maintenance to prevent criticalities.

Similar instabilities appear in a variety of other infrastructures. Communication systems, such as the telephone network, are one example. But probably the most sensitive is the transportation network: streets, highways, and railways connecting cities, the web of boats transporting fuel and other goods, and above all the airport web. Planes transport billions of passengers and tonnes of goods every year. A minimal malfunction in such an infrastructure has major consequences: Eurocontrol (the European organization for the safety of air navigation) has estimated that delays in flights cost European countries up to €200 billion in 1999 alone. In a globalized world, transportation networks are similar to circulatory systems in living beings.

A net as large as the world

In October 1969, a message travelled for the first time from one computer to another, through a telephone line. Two university labs in California were at the ends of the line. After a few letters, the message broke down, but the connection was established: Arpanet, the grandfather of the Internet, was born. The idea of a network of computers was around during the previous decade. At the end of the fifties, ARPA (the US Advanced Research Project Agency) asked engineer Paul Baran to design a communication structure able to resist an attack. In particular, the whole system had to continue working even under an attack destroying part of it. Baran duly designed a distributed system with such characteristics—but a change in strategy locked his pioneering studies into a drawer. However, in the sixties, some universities asked ARPA to finance a similar project for different purposes. The academic institutions were keen to interconnect their computers in order to aggregate their computational power.

The 1969 Arpanet connected UCLA (University of California Los Angeles) and the SRI (Stanford Research Institute). Two years later, the number of nodes was over forty, including some companies and other universities. This structure was so successful that in the seventies similar networks appeared in other parts of the world, created by particle physicists, astronomers, companies: Hepnet, Span, Telenet, etc. If at the beginning the problem was to connect the computers, this moved on how to connect networks. *Internetworking* became the motto of many computer scientists. At the end of the seventies, engineer Robert Kahn and mathematician Vinton Cerf developed the TCP/IP: whatever the internal structure of networks, this software allows them to talk to each other. This code was put in the public domain and based on the concept of *open architecture*. Finally, in the eighties, the *TCP/IP transition* was fully achieved, bringing to creation the Internet, the 'network of networks'.

Such a structure is probably the human artefact that best embodies the idea of networks. A computer connected to the Internet becomes one of many *hosts*. If we want to deliver an email to a specific place, we do not need to be directly connected with that destination. From origin to our target, information travels along *routers*, devices responsible for transmitting packets of data. A large series of connections keeps the structure linked: optical fibres, telephone lines, satellite connections, etc. Since nobody plans where hosts and connections are added, the overall structure of the Internet is not recorded. Actually, mapping at host level is practically impossible. We can have a rough representation of it only at router level. In this case, the nodes of these networks are the routers, and the edges are their connections. We can coarse-grain the structure even more, grouping routers into *autonomous systems*. These groups are autonomously administrated domains that usually correspond to Internet Service Providers (ISPs) and other organizations.

The great success of the Internet is due to the exceptional experience it provides. Watching television is a one-direction, one-medium, passive experience. The Internet is not so. People can navigate an infinite series of documents, use different media, exchange information, and talk to each other. Unlike traditional communication technologies such as telephone, radio, or television, the Internet does not have a specific purpose. Rather, it is a mutant artefact able to host infinite applications. Actually, the Internet is only a physical infrastructure that supports services. One of the most successful of these is the World Wide Web (WWW). This is an enormous set of electronic *documents* recorded in the devices that make up the Internet and connected by *hyperlinks* that allow navigation between them. The pattern is somewhat similar to the body of scientific literature, made of articles, books, patents, etc. connected by citations.

The idea of the WWW was born at CERN (European Organization for Nuclear Research). Physicist Tim Berners-Lee (later joined by computer scientist Robert Cailliau) put forward the proposal for it in 1989. Berners-Lee designed a system that allowed scientists to access, through their own computers, the enormous amount of data produced by particle physics experiments. The software to make this system work was not patented but rather released in the public domain. This decision—as in the case of TCP/IP—proved to be very significant. Right from the beginning, thousands (and then even greater numbers) of users tried it, improved it, and created web pages and services. In just a few years, the Web became World Wide. Its magnitude is unknown, since none of the search engines that explore the web (such as Google or Yahoo!) is able to archive all web pages. After all, this would make little sense: several websites are capable of producing new pages upon request. A 2005 estimate put the content of the whole WWW (for static pages) as equivalent to 200 *terabytes* of information. At the time, this was about ten times the size of the US Congress Library. Undoubtedly today's figure would be greater by orders of magnitude, since the growth of the WWW is exponential.

Cyberspaces

On 11 September 2001, the infrastructures of New York City experienced a 'network catastrophe', parallel and related to the human tragedy that was taking place that same day. Soon after two hijacked airliners crashed into the World Trade Center complex, a surge in phone calls was registered. People were trying to communicate with friends, and rescue personnel with colleagues. Cell networks were quickly overloaded and people lined up at payphones in Manhattan. The attack damaged Verizon's central office, interrupting 200,000 lines. AT&T infrastructures, some of them housed in the basement of the World Trade Center were destroyed as well. When calls failed, many people turned to the Internet. But wireless service was also impacted. The impact on the economy extended well beyond the crash sites. It took six days before the New York Stock Exchange could return to work. Months passed before services were restored to near pre-disaster levels.

The terrorist attacks showed that there is almost no network that stands alone. Physical and virtual infrastructures are embedded in a common *cyberspace*, where energy, information, transport, communication, etc. are provided. Power grids support the Internet, that hosts the WWW, that in its turn enables email services, social networks, information websites, and file-sharing systems. Many activities, including flight control, bank programs, emergency systems, and commercial services also depend on the WWW. A collapse at one level of cyberspace can affect the other layers, often in a quite unpredictable way.

The interconnection of multiple networks is common in many other situations. For instance, the liquidity shock that affected the economy in 2008 rapidly spread to many other economic networks. Similarly, social networks show this feature in many aspects. One interesting example is that of real-world friendships as compared with virtual social network contacts: non-trivial feedbacks exist between the two networks. Cells are a miniature cyberspace: genetic regulatory networks, protein interaction networks, and metabolic networks are nested into one another; for this reason, some scientists have proposed fusing the concepts of genome, proteome, metabolome, etc. into the comprehensive idea of *interactome*. Finally, ecosystems can be seen as sets of interacting networks; for example, both antagonistic and mutualistic networks play a role in determining how species will thrive. In all these cases, networks provide useful maps to disentangle complex and interwoven systems.

Chapter 4
Connected and close

One world

On 4 November 2006, the disconnection of a single power line in north-west Germany triggered an avalanche of blackouts as far away as Portugal. One expects that such a small disconnection could have effects at the regional level, or at most within the national power grid. But electrical networks have become more and more integrated (currently, they form one large system at continental level) and then more fragile. The same has happened in other infrastructures. For example, almost all airports are interconnected: today it is possible to reach practically every destination, starting from any point of origin, with a limited number of stops on the way. The Internet is fully connected as well, since it has grown out of the integration of smaller networks.

Networks in nature and society may very well be fragmented in separated parts. In the cell, some groups of chemicals interact only with each other and with nothing else. In ecosystems, certain groups of species establish small foodwebs, without any connection to external species. In social systems, certain human groups may be totally separated from others. However, such disconnected groups, or *components*, are a strikingly small minority. In all networks, almost all the elements of the systems take part in one large connected structure, called a *giant connected component*.

For example, in one experiment based on mental associations between words, scientists found that 96 per cent of the terms ended up in one large group. One could find a path between any two expressions in this group, even such different terms as 'volcano' and 'stomach': the chain of associations drawn by the people participating in the experiment was 'volcano', 'Hawaii', 'relax', 'comfort', 'pain', and 'stomach'. In general, the giant connected component includes not less than 90 to 95 per cent of the system in almost all networks. A few interesting consequences of this fact can be listed. In sexual interaction networks, past and current intercourse can connect us with individuals we would never imagine or desire to be related to. In the network of collaboration between scientists, a large cooperation structure appears, omitting only a few lonely players. In company boards, interlocking grants a kind of connectedness that encompasses the vast majority of companies. Finally, in foodwebs, pollutants introduced in one species at a given location are brought by foodchains to apparently unrelated species as far away as on the opposite side of the planet.

Living in a big, connected world does not always imply that every node can be reached from every other. As with ordinary roads, where it is crucial to know if they are one-way or not, we must know if edges are directed. In a directed network, the existence of a path from one node to another does not guarantee that the journey can be made in the opposite direction. Wolves eat sheep, and sheep eat grass, but grass does not eat sheep, nor do sheep eat wolves. This restriction creates a complicated architecture within the giant connected component (Figure 6). For example, according to an estimate made in 1999, more than 90 per cent of the WWW is composed of pages connected to each other, if the direction of edges is ignored. However, if we take direction into account, the proportion of nodes mutually reachable is only 24 per cent, the

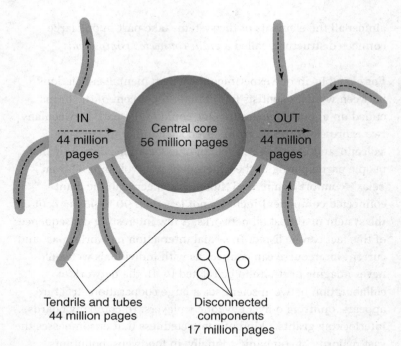

6. Networks with directed links, like the WWW, display a bow-tie structure: a central giant strongly connected component, an in-component, and an out-component, as well as minor structures (tubes, tendrils, and a few disconnected components). Data in the picture are from 1999.

giant strongly connected component. The rest is divided between an *in-component* and the *out-component*: the first is made up of pages with paths pointing to the giant strongly connected component, and the second of nodes that receive links coming out from it (the picture is completed by minor structures called *tubes* and *tendrils*). This characteristic structure gives the giant connected component of the WWW a characteristic 'bow-tie' shape. This complicated structure is not specific to the WWW, but appears, with different components, in all directed networks.

The existence of the giant connected component—either with a bow-tie structure or not—is a remarkable feature. For example, the giant strongly connected component of the WWW, although being less than one-third of the system, still encompassed 56 million pages in 1999. An obvious explanation of this feature would arise if networks were very *dense*—that is, if they had many redundant links, sufficient to connect every node with almost every other. Yet usually this is not the case. On the contrary, most networks are *sparse*, i.e. they tend to be quite frugal in connections. Take, for example, the airport network: the personal experience of every frequent traveller shows that direct flights are not that common, and intermediate stops are necessary to reach several destinations; thousands of airports are active, but each city is connected to less than 20 other cities, on average. The same happens in most networks. A measure of this is given by the mean number of connection of their nodes, that is, their *average degree*. We create many web pages every day, but the average number of links is about ten. Hundreds of thousands of routers connect the Internet, but each of them connects to fewer than three others, on average. Finally, in a sample of more than 50,000 physicists, the average number of collaborators was found to be about nine. In most real-world networks, there may be elements with many connections (and indeed, there are), but in general the graphs are not dense: on the contrary, they are said to be *sparse*.

This puzzling contradiction—a sparse network can still be very well connected—had already attracted the attention of the Hungarian mathematicians we encountered in Chapter 2, Paul Erdős and Alfréd Rényi. They tackled it by producing different realizations of their random graph. In each of them, they changed the density of edges. They started with a very low density: less than one edge per node. It is natural to expect that, as the density increases, more and more nodes will be connected to each other. But what Erdős and Rényi found instead was a quite abrupt transition: several disconnected components coalesced suddenly into a large one, encompassing almost all the nodes. The sudden change happened at one specific critical density: when the average number of links per node (i.e. the average degree) was greater than one, then the giant connected component suddenly appeared. This result implies that networks display a very special kind of economy, intrinsic to their disordered structure: a small number of edges, even randomly distributed between nodes, is enough to generate a large structure that absorbs almost all the elements.

So close

In early 1994, three students at Albright College—Craig Fass, Brian Turtle, and Mike Ginelli—were watching television during a heavy snowstorm. According to their own reports, they noticed that Kevin Bacon, whose next movie was announced on television, was present in many different movies. So they began to speculate about the large number of actors that had starred with him in those movies. The idea that Bacon was a kind of 'center of the entertainment universe' started to spread, became famous, and a web page based on it even appeared, the *Oracle of Kevin Bacon*: this search engine provides the relation between Bacon and any actor one may search for. Remarkably, if one writes the name of a Spanish actor from old commercial movies, such as Paco Martínez Soria, the oracle finds a quite close relation: Martínez acted in *Veraneo en España* with Luís Induni; the latter acted in *Il Bianco, il Giallo e il Nero* with Eli Wallach; and Wallach acted in *Mystic River* with Kevin Bacon. Such short chains of collaborations with Bacon are found for almost any actor one may think about.

This surprising feature brings to mind a game played by scientists. Paul Erdős, the expert of random graphs, was one of the 20th century's leading mathematicians. Scientists assign themselves, as a badge of honour, a measure of their collaboration with him: those that have co-authored a paper with him are said to have an Erdős number 1; co-authors of his co-authors have Erdős number 2; co-authors of co-authors of Erdős' co-authors have Erdős number 3; and so on. However, only the lowest Erdős numbers are a real source of pride: more than 500 scientists were direct co-authors of Erdős'; and a few thousands have collaborated with this nucleus of co-authors. Finally, tens of thousands people have Erdős number 3 (G. C. is one such; M. C. has Erdős number 4), so it is not exactly something of special merit. Almost no scientist in any field has an Erdős number greater than 13.

Contrary to what one may believe, these notable results are nothing specific to Kevin Bacon or Paul Erdős. The former is not the centre of the entertainment world; nor is the second the hub of mathematics. If the same calculations are repeated with any other actor or scientist, similar results are found: very short chains connect apparently remote individuals. This fact gives an interesting insight into a quite common *cocktail party experience*: you are talking with a stranger and suddenly discover that he or she is your wife's schoolmate, or your brother's tennis partner, or your friend's neighbour. This discovery is usually hailed as a surprise ('It's a small world!'), but it may not be so unusual. Social systems seem to be very tightly connected: in a large enough group of strangers, it is not unlikely to find pairs of people with quite short chains of relations connecting them.

Six degrees of separation

In 1967, American psychologist Stanley Milgram undertook a set of memorable experiments. With the collaboration of Jeffrey Travers, he sent tens of letters to citizens in the Midwest (Kansas and Nebraska), chosen at random. In the message, he asked them to forward the letter to a person in Massachusetts (either, the wife of a divinity student at Cambridge or a stockbroker in Boston), the address of whom he did not provide. In case they did not know the recipient, he suggested they forward the letter to somebody they knew, that might for some reason be 'close' to the recipients. For every forwarded letter, another one had to be sent to Milgram himself, so that he could follow the path of the messages. In a country of hundreds of millions, finding somebody by essentially a word-of-mouth procedure seems impossible. However, after a few days, the recipients started to receive the first letters. These had passed through only one intermediary. A few weeks later, when the experiment was declared finished, about a third of the letters had arrived at their destination: none had been sent more than ten times and, on average, the number of mailings was six.

The experiment sparked enthusiasm in the scientific community. In the 1950s, Manfred Kochen, a mathematician, and Ithiel de Sola Pool, a political scientist, had speculated that humans may be much 'closer' to each other than expected. They asked: If two persons are selected at random from a population, what are the chances that they would know each other? More generally, how long is the chain of acquaintanceship that is required to link them? In a highly circulated paper, which they eventually published in 1978, they proposed a mathematical model suggesting that, in a population such as that of the United States, an unexpectedly large fraction of pairs could be linked by chains with just a few intermediaries. Milgram's experiment was an empirical test of their intuition. This finding had a strong impact even beyond the academic world. The expression *six degrees of separation*, referring to the number found in Milgram's experiment, became the popular formulation of his results. In 1990, playwright John Guare used it as the title of a comedy in which a charismatic character dodges people, saying that he is the son of actor Sidney Poitier. Here is how Guare expressed Milgram's results:

> I read somewhere that everybody on this planet is separated by only six other people. Six degrees of separation. Between us and everybody else on this planet. The President of the United States. A gondolier in Venice. Fill in the names. [...] It's not just big names. It's anyone. A native in a rain forest. A Tierra del Fuegan. An Eskimo. I am bound to everyone on this planet by a trail of six people.

Small worlds

Iloveyou was one of the most contagious computer viruses ever. After appearing on May 2000, it hit tens of millions of computers all over the world, producing billions of euros of damage, mainly the costs of eradicating it. *Iloveyou* arrived in an email, disguised in an attachment that looked like a love letter. When the attachment was opened, the virus infected the computer, and forwarded itself to the email addresses contained in the address book of that computer. After just a few replications of this kind, the virus reached an enormous number of devices. As in the social networks described in the previous paragraph, one could say about the computer network where viruses spread: 'It's a small world!' A large number of computers are reached through just a few connections. Computers apparently far from each other turn out to be connected by short chains of links.

In reality, this *small-world property*, which is the main result of Milgram's experiment, is present in all networks. The Internet is composed of hundreds of thousands of routers, but just about ten 'jumps' are enough to bring an information packet from one of them to any other. Thousands of kilometers may stand between them, but that is not the distance that matters: what matters is the number of connections to be crossed, and this is always very small. That is why information travels across the planet at an extraordinary velocity. Another example is the WWW: it is composed of billions of pages, but scientists have found that about twenty mouse 'clicks' are enough, on average, to navigate between any two of them. There are fewer than three 'degrees of separation', on average, between any pair of neurons in the brain of a *C. elegans*. In the import–export network that interconnects the countries of the world, it is impossible to find two states separated by more than two links. The list of examples could go on with many other cases.

The small-world property consists of the fact that the average distance between any two nodes (measured as the shortest path that connects them) is very small. Given a node in a network (say, Paul Erdős in the co-authorship network), few nodes are very close to it (direct co-authors) and few are far from it (scientists with very high Erdős numbers): the majority are at the average—and very short—distance. This holds for all networks: starting from one specific node, almost all the nodes are at very few steps from it; the number of nodes within a certain distance increases exponentially fast with the distance. Another way of explaining the same phenomenon (the way scientists usually spell it out) is the following: even if we add many nodes to a network, the average distance will not increase much; one has to increase the size of a network by several orders of magnitude to notice that the paths to new nodes are (just a little) longer.

The small-world property is crucial to many network phenomena. The short synaptic distance in the neocortex may be crucial to its functioning: some studies suggest that neurodegenerative illnesses such as Alzheimer's imply a massive loss of the small-world property in the brain. Short distances in the sexual relations networks suggest that the concept of risk group in sexually transmitted diseases has to be interpreted carefully: virtually everybody is at a very short distance from somebody infected. This capability of networks to spread viral agents efficiently can also have constructive applications. One of the first examples of *viral marketing* strategy was the worldwide diffusion of the *Hotmail* email service, launched in 1996. People purchasing a free Hotmail address agreed to host in their messages a link that allowed recipients to open a free address in their turn. Hotmail had one of the fastest increases for a communication company, attacting tens of millions of users, partially due to this strategy that cleverly exploited the small-world nature of the email network.

Shortcuts

The small-world property is something intrinsic to networks. Even the completely random Erdős–Renyi graphs show this feature. By contrast, regular grids do not display it. If the Internet was a chessboard-like lattice, the average distance between two routers would be of the order of 1,000 jumps, and the Net would be much slower: no quick web browsing, no instant emails. If the network of scientific collaborations was a grid, Paul Erdős would have a moderate number of co-authors; these would have a larger, but still moderate, number of collaborators, etc.: the number of individuals within a certain distance would not grow exponentially but much more slowly. If the neurons network was a lattice, increasing the number of neurons (due to the physical growth of the brain, for example) would increase remarkably the average communication distance within the neocortex: paradoxically, growth would make people less smart (something young readers could agree with).

What makes a network different from a grid? What factor is responsible for the appearance of the small-world property in webs, while lattices lack it? In 1998, physicist Duncan Watts and mathematician Steven Strogatz tried to answer these questions. They started by considering a very simple, regular structure. It was a circle of nodes, and each of them was connected to its first- and second-nearest neighbours (Figure 7 left). This structure may represent a group of remote villages, each of them interchanging goods with the neighbouring villages, and occasionally with the neighbours of their neighbours. In such a regular structure, products can travel a long way from producers to consumers in faraway villages.

Watts and Strogatz then allowed for something very similar to opening a path between two distant villages. In practice, they cut

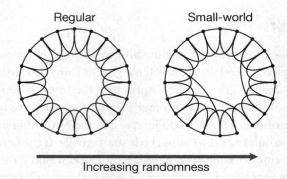

7. **In the small-world network model a regular lattice is transformed into a network by introducing disorder, and correspondingly the distances between nodes drop: the small-world property arises**

one of the links in the initial structure and rewired it with another node, chosen at random. Suddenly, the inhabitants of one village can interchange goods with a previously remote area, instead of their neighbours. Still, on the large scale, just a few villages are affected by this change, and several areas of the circle remain far from each other. One can see this by computing the average distance between nodes after the rewiring: it is just moderately reduced by the new shortcut. At this point the two scientists allowed for more 'paths' (Figure 7 right). After every rewiring, they computed the average distance: remarkably, they found that, after rewiring just a few links, the distance dropped abruptly. A small number of shortcuts is enough to bring all the elements of the system much closer to each other. The key ingredient that transforms a structure of connections into a small world is the presence of a little disorder. No real network is an ordered array of elements. On the contrary, there are always connections 'out of place'. It is precisely thanks to these connections that networks are small worlds.

These shortcuts are easy to identify in some nets. For example, in 1858, the first transatlantic telegraph cable connected Europe and

America. This wonder of thousands of kilometres and hundreds of tonnes is one of the marvels of the oceans visited by the submarine in Jules Verne's *Twenty Thousand Leagues Under the Sea*. Today, several transoceanic cables grant immediate spreading of information throughout the world. In word networks, polysemy is one main source of shortcuts. For example, the word 'pupil' connects the two semantic areas of teaching on the one hand (pupil as student), and vision on the other (pupil as the body organ). In social networks, Granovetter's concept of *weak ties*—links that connect unrelated groups—could be at least partially matched with Watts and Strogatz's shortcuts.

Shortcuts are responsible for the small-world property in many other situations. However, every now and then we can find another possible explanation. For example, the remarkably small distances in the world trade web are due to the fact that this is one of the few highly dense (non-sparse) networks: the average number of partners of a country is typically comparable with the total number of countries represented in this network, something that suggests that each one has interchanges with a large majority of the others. In foodwebs, other mechanisms favour the small-world property. Basal species get energy and matter from sun and environment, but their predators extract just 10 per cent, on average, of the resources contained in them, and the same happens at every successive step of predation: if foodchains were too long, top predators would not extract enough from them to survive.

Whatever its origin, the small-world property is a crucial feature to be taken into account when systems have the structure of a graph. The network approach provides a striking vision of such systems: in the first place, their elements are part of one big world, where almost every node has a path of connections with every other; in the second place, these paths are extremely short. This interwoven structure is essential to understanding a broad range of phenomena, from Aids to blackouts to information spreading.

Chapter 5
Superconnectors

Hubs

In his memorable experiments on the 'six degrees of separation', Stanley Milgram made an observation that was going to be fully understood only much later. In one of his experiments, the US psychologist asked random citizens of Nebraska to forward a letter to a stockbroker in Massachusetts. If they did not know the recipient, they should send the letter to somebody that they believed to be closer to him. Beyond the fact that a large part of the letters arrived in an average of just six steps, Milgram observed that a quarter of them were delivered to the recipient from the very same source: a clothing merchant and friend of the stockbroker, whom Milgram calls Mr Jacobs. This result was quite mysterious: how was it that so many paths leading to the stockbroker passed through this person?

Frequent air travellers are familiar with a very similar phenomenon. Airports like Heathrow, Frankfurt or JFK in New York are well known to globetrotters: whatever the destination, it is quite likely that flights stop over at those airports. Airline magazines often carry a map of the world, crossed by long lines that show their routes: many of them end up at or pass through places like London, Frankfurt or New York. Airports like these are called *hubs* and carry a large portion of the overall traffic.

It's easy to argue that the role of Mr Jacobs in the social network is the same as the big airports in the air traffic network. Probably, Jacobs was a social relations hub: his many contacts connected him to several people, so it was natural that many letters passed through his hands.

Another striking observation made by Milgram is that a large part of the remaining letters arrived from just two other people: Mr Jones and Mr Brown. Using the air traffic metaphor, these two people were most probably 'average-size airports' (like Madrid or Milan) of the social network. The remaining letters, that didn't came from Jacobs, Jones, or Brown, passed through smaller 'airports' (like Girona or Olbia) of the social network.

The presence of these hubs is not specific to the stockbroker's social network, or to the airport network. Many other systems, when represented as graphs, show similar highly connected vertices, or *superconnectors*. In many networks, one can see a 'winner takes all' tendency: a few nodes attract the majority of connections, and the large remainder of nodes have to share the remaining links.
A modern analysis has shown that Mozart's Don Giovanni (who seduced 2,065 women, according to Da Ponte's libretto: '... 640 in Italy, 231 in Germany, 100 in France, 91 in Turkey, but in Spain they are already 1,003 ...') was not an exaggeration: the most connected individuals in sexual interaction networks can reach thousands of intercourses. In some datasets, some of these are people involved in the sex trade. Naturally, these highly connected individuals are those most needing protection against sexually transmitted diseases. Another example of superconnector was found early after the September 11 terrorist attack on New York: management consultant Valdis Krebs drew a simple map of the social networks of the terrorists and found that Mohammad Atta, one of the leaders of the conspiracy, was the most connected node, that is, a hub. In scientific collaboration networks, one can also find

key figures that cooperate with a large number of colleagues: Paul Erdős was one of them.

Superconnectors are present in many kinds of network, not only in social ones. Some routers in the Internet have thousands of connections: that is, thousands more than the average router, which has just a few links. *Meet-me rooms* are large facilities—usually buildings full of cables—where hundreds of Internet Service Providers can link to each other: the failure of one of these facilities can leave entire areas as (as large as a state) without Internet connection. Websites of large newspapers attract an enormous number of links from other websites, blogs, and social networks. In foodwebs, top species predate a large quantity of other species. Finally in words' networks, the hubs are the ambiguous or polysemic words: such as 'arms', which in English refers both to bodily extremities and weapons, and thus connects to a larger semantic or synonym field.

Hubs are also present in the networks inside the cell. In the genetic regulatory network, a single gene can control the expression of a large portion of the rest of the genome: in one bacterium (the *Caulobacter crescentus*), one single regulatory factor (the CtrA) controls 26 per cent of the cell cycle-regulated genes. The p53 molecule is a superconnector of the protein interaction network: the gene associated to this protein is a powerful tumor suppressor, and it is mutated in a large range of tumors. A clear hub of the metabolic network is the Atp molecule (adenosine triphosphate): this plays the role of energy carrier for a large number of biochemical reactions.

Giants, dwarfs, and networks

And there came out from the camp of the Philistines a champion named Goliath of Gath, whose height was six cubits and a span. [...] He was armed with a coat of mail, and the weight of the coat was five thousand shekels of bronze [...]

> The shaft of his spear was like a weaver's beam, and his spear's head weighed six hundred shekels of iron. (1 Samuel 17:4—7)

According to the biblical book of Samuel, Israelites have to wait 40 days before someone dared to face somebody as imposing as Goliath: then enters David, a brave and reckless boy, who ends up defeating the enemy. This was not a common adversary: the 'six cubits and a span' height corresponds to about 3 metres, and the 'five thousand shekels of bronze' weight of his coat of mail would be between 60 and 90 kg, according to historians' estimates.

The conversion of ancient measures into modern ones is not precise; moreover, the biblical narration is most probably symbolic. However, Goliath's size is not completely unlikely. According to the *Guinness Book of Records*, the tallest person ever recorded, an American called Robert Wadlow, was 2.75 metres tall. Unlike Goliath, who has special armour and a spear suited to his size, the exceptionally tall are usually surrounded by objects too small for them: chairs are uncomfortable, ceilings are too low, and they need to wear specially tailored shoes and clothing.

The root of their problems is that body size is a *homogeneous* magnitude. People entering a cinema are of different sizes, but all the seats are identical: some people find them large, others small, but in general they are reasonably comfortable. Body size does not vary a lot from the average. Very tall (or very short) people are exceptionally rare, and the taller (or shorter) they are, the rarer. Almost everybody knows somebody about 1.90 metres tall, but few know a 2-metre tall person and almost nobody a 2.5-metre one. People are also homogeneous with respect to other features. For example, IQ tests yield results close to the average most of the times, and deviations—both upwards and downwards—are rare. Behaviours can be quite homogeneous, too. For example, drivers can be more or less reckless, but most of the time the speed measured on highways conforms quite closely to the average.

However, homogeneity is not always the rule. For example, the number of friends a person can have is extremely variable. Robert Wadlow was 'only' five times taller than the shortest person ever, Chandra Bahadur Dangi, 55 cm tall, according to the *Guinness Book of Records*. By contrast, the friendliest people (the hubs of social networks) can have tens or hundreds of friends more than the extremely shy, who interact with very few individuals. If contacts in virtual social networks are taken as a proxy of the number of friends a person has, then the hubs of these networks have hundreds more friends than the less connected people. While height is a homogeneous magnitude, the number of social connection is a *heterogeneous* one.

If the height of people reflected the number of their social connections, somebody as tall as Wadlow would not enter any record book. There would be people hundreds of times taller than the shorter ones: 'social giants' more than 2 km tall would walk the streets. Even more interestingly, these giants would not be astounding exceptions in a generally short population. All the intermediate heights between dwarf and giant would be represented by some individuals: naturally, the greater the height, the fewer the people of that height; however, the number of tall people in this imaginary world would not diminish as quickly in terms of height as in the real world. In other words, the taller, the rarer: but not as rare as in the real world.

The business of seat manufacturers would be much more difficult in this world, because there would be no way to build a seat that would fit every body size. In the real world, if we want to manufacture seats, or analyse IQ tests, or predict the duration of a road trip, we take into account average height, IQ, or speed. But in order to understand social relations, the very concept of average may be useless. Body size, IQ, road speed, and other magnitudes have a *characteristic scale*: that is, an average value that in the large majority of cases is a rough predictor of the actual value that one will find. In contrast, social relations do not have this scale. If you knock on the door of an unknown neighbour, you expect to see a person whose height is within a certain reasonable range, and most of the time your guess will be accurate. But it is almost impossible to guess in advance whether that person has many or few friends, and how many. The average number of relationships in a town just gives an idea about whether the social network of that place is more or less dense. But it does not allow us to make any reasonable prediction about each single person.

A system with this feature is said to be *scale-free* or *scale-invariant*, in the sense that it does not have a characteristic scale. This can be rephrased by saying that the individual *fluctuations* with respect to the average are too large for us to make a correct prediction.

Fat tails

In general, a network with heterogeneous connectivity has a set of clear hubs. When a graph is small, it is easy to find whether its connectivity is homogeneous or heterogeneous (Figure 8). In the first case, all the nodes have more or less the same connectivity, while in the latter it is easy to spot a few hubs. But when the network to be studied is very big (like the Internet, the Web, metabolic networks, and many others) things are not so easy.

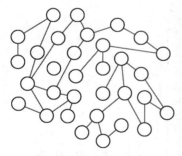
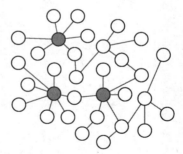

Homogeneous network   Heterogeneous network

8. **A homogeneous network (left), where all nodes have more or less the same degree, compared with a heterogeneous one (right), where highly connected nodes (hubs) are present**

Fortunately, mathematics provides a way to find whether a magnitude is heterogeneous or homogeneous.

Let us start with a homogeneous magnitude, such as people's height. In order to study the height of the students of a class, one can do as follows. First, make a row with those 1.50 to 1.55 metres tall: there will be a few of them. Then, make a parallel row with those 1.55 to 1.60: there will be some more and the row will be a little longer. Follow those with 1.60 to 1.65: more people will be in that row. Increasing by 5 centimetres in every row (Figure 9 left). At the end, the profile of the rows will have the shape of a *bell curve*: the number of students increases as height increases, then reaches the top around the average, and then starts to fall. The very tall and the very short are rare and the majority are in the middle. This curve provides the distribution of heights of the students.

Now, let us consider the number of social contacts of those same students. Now the rows correspond to those with 0 to 20 friends, 20 to 40, 40 to 60, and so on. The outcome of this procedure provides the distribution of the connectivity of the nodes of the social network, that is, the *degree distribution* of the graph. The resulting picture is very different from the case of heights (Figure 9 right). First of all, there will be many more rows, since

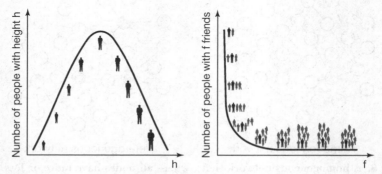

9. **Height is a homogeneous magnitude, distributed according to a bell curve (left), while the number of friends is a heterogeneous magnitude, distributed according to a power law (right)**

there will be people with hundreds or thousands of friends. The majority of people will have some tens of contacts per person, but the resulting distribution will have a 'fat tail'. In other words, the distribution will be very skewed to the right, with a long or 'heavy tail'. Mathematically speaking, the shape of the degree distribution is well described by a curve called *power law*.

In homogeneous networks, the degree distribution is a bell curve, similar to that of height, while in heterogeneous networks, it is a power law, similar to that of friendships. The power law implies that there are many more hubs (and much more connected) in heterogeneous networks than in homogeneous ones. Moreover, hubs are not isolated exceptions: there is a full hierarchy of nodes, each of them being a hub compared with the less connected ones. Take once again the comparison of height and friendships. Probably there are several million people in the world who are 1.50 metres tall; however, if we double that height (3.00 metres), the number of people this tall is much smaller, most probably zero. On the other hand, tens of millions of people have, say, 20 friends in their social network. If we double that number (40 friends), there will be fewer (say a quarter of those with 20), but still millions. We can double the number many times, and the number will be reduced by about a quarter at every step (the actual rate of reduction depends on the slope of the power law). This explains, for example, the role of Mr Jones and Mr Brown in Milgram's experiment: while Jacobs is the largest hub of the stockbroker's social network, Jones and Brown are smaller hubs, but still well connected.

Looking at the degree distribution is the best way to check if a network is heterogeneous or not: if the distribution is fat tailed, then the network will have hubs and heterogeneity. A mathematically perfect power law is never found, because this would imply the existence of hubs with an infinite number of connections. However, no real network is infinitely big: this is why the fat tail of the degree distribution always has a cut-off at a maximum value for the degree. Indeed, the size of hubs can be limited by various costs of accumulating connections: for example, neurons cannot accumulate an arbitrary amount of connections, because of their physical structure. In professional collaboration networks, time plays a role: connections cannot be accumulated indefinitely, because at a certain moment the career (or life) of an individual comes to an end. All these and other factors are reflected in the shape of the degree distribution. Nonetheless, a strongly skewed, fat-tailed distribution is a clear signal of heterogeneity, even if it is never a perfect power law.

One has to be careful in interpreting what hubs and fat tails mean. For example, some anthropologists believe that a magnitude called *Dunbar number* limits the number of social relations. According to this hypothesis the number of stable social bonds cannot increase much above 150. Anthropologist Robin Dunbar put forward this hypothesis in 1992 after finding evidence that the size of a part of the brain's cortex of primates and humans may be related to that of their social groups. If this is true, what is the explanation of the hubs with thousands of connections found in many social networks? Some scientists think that they are instances of the *pizza delivery guy problem*. A pizza delivery guy receives many phone calls on his mobile, but just a tiny fraction of them come from real friends; the rest are clients. According to this concept, the majority of the links represented in the fat tail of the distribution would be fictitious in social networks. However, this depends on exactly what problem one wants to study. For example, if the pizza delivery guy catches flu, epidemiologists only care about how many people (friends or not) have been in contact with him.

On the other hand, not all networks are necessarily heterogeneous. While the small-world property is something intrinsic to networked structures, hubs are not present in all kind of networks. For example, power grids usually have very few of them. The same holds for some foodwebs, the neuronal network of the *C. elegans*, and the World Trade Web.

Finally, an interesting case is found in some directed networks, as in the case of, most genetic regulatory networks. If gene A regulates gene B, an arrow is drawn from A to B, but not necessarily from B to A. The *out-degree distribution* (that is, the distribution of the number of outcoming arrows) is usually fat tailed: that is, a few genes regulate large portions of the genome. However, the *in-degree distribution* (that of the number of incoming arrows) is much more homogeneous: just a few other genes regulate a single one. Heterogeneity is widespread in many networks, but when we approach an unknown system we must not take it for granted until we have checked it.

The signature of self-organization

Heterogeneity and the lack of characteristic scale may very well be a sign of disorder. One could reason as follows. Many networks (like the Internet or social networks) have grown without any blueprint or supervision. As a consequence, every node in the network follows its own criteria and performs completely different and uncoordinated behaviours. These are so disordered that they can be easily assimilated to an overall random process. As a consequence, random graphs should be good models for these networks. This line of reasoning seems to work, but some problems appear after deeper inspection. The most noticeable is that random networks are not heterogeneous at all. On the contrary, their degree distribution is bell shaped, suggesting that all nodes have more or less the same degree. The process of connecting pairs at random is such that every node ends up with more or less the same degree. More precisely, the degree has a characteristic scale, with small fluctuations around the average. In contrast with many real-world networks, hubs are not present in random networks.

A consequence of this is that, while random networks are small worlds, heterogeneous ones are *ultra-small worlds*. That is, the distance between their vertices is relatively smaller than in their random counterparts. If one takes a random network and adds a certain number of hubs to it (thus making it more heterogeneous), then the distance becomes smaller. Conversely, if one takes a heterogeneous network and randomizes it (that is, builds a network with the same number of nodes and edges, but with edges distributed at random), then hubs disappear, and the average distance becomes larger. This shows that hubs are responsible for the majority of the connectivity of these networks: a large portion of the connections arise precisely from this small number of superconnected nodes.

More importantly, the fact that random networks are homogeneous means that the equivalence between heterogeneity and disorder is flawed. A disordered process such as the one described by random networks does not yield the heterogeneous connectivity found in many real-world networks. On the contrary, heterogeneity may arise from the exact opposite: that is, from some kind of regular, ordered behaviour.

This is rather puzzling, since many networks are not the result of a blueprint, nor do they evolve under tight top-down supervision. Few networks, like the electrical or road ones, are controlled by political and technical authorities, but most of them are unsupervised. The Internet, for example, is controlled by network administrators at the local level, and is also constrained by technical, economic, and geographic features. However, its large-scale structure is largely unplanned: the Internet is very similar to a global-scale experiment, where nobody draws the overall structure, which is the result of the actions of innumerable agents. Biological networks are an even clearer example: there is no designer, only the tinkering effects of evolution. With respect to social networks, politics, money, religion, language, and culture influence the relations between individuals, but when spaces of freedom are available, the shaping of these networks is not strictly planned. In all these cases, the overall organization of the systems emerges from the collective action of its elements, a bottom-up process of *self-organization*. This process may explain why many networks, even without being blueprinted, still display a remarkable signature of order like heterogeneity.

Heterogeneity is not exclusive to networked systems. For example, the intensity of earthquakes has a fat-tailed distribution: if one plots the frequency of earthquakes versus their intensity, a nice power law emerges. The 'average earthquake' does not exist; there is wide variability, from imperceptible vibration to large-scale catastrophes. Another example is the size of cities: this ranges from the largest Chinese megalopolis to small towns in the countryside of Tuscany. Another example is the distribution of income: At the beginning of the 20th century, economist Vilfredo Pareto showed that 80 per cent of Italian land was in the hands of 20 per cent of the population; various levels of skew of this kind are present in all economies.

All these examples share with networks one basic feature: they are the outcome of a complex, largely unsupervised process. Heterogeneity is not equivalent to randomness. On the contrary, it can be the signature of a hidden order, not imposed by a top-down project, but generated by the elements of the system. The presence of this feature in widely different networks suggests that some common underlying mechanism may be at work in many of them. Understanding the origin of this self-organized order is one of the central challenges of the science of networks.

Chapter 6
Emergence of networks

Everlasting change

By the nineties, the Internet was mostly an unknown land. Although it was already a critical infrastructure for communication, trading, and transportation, nobody had a clear idea about its overall architecture. Administrators controlled local networks, but had no clue about the large-scale structure of the net. Moreover, the increase had been explosive: from some tens of machines at the beginning of the seventies, to tens of millions, with even larger prospects of growth. By the end of the decade, organizations like the computer company Compaq and the Cooperative Association for Internet Data Analysis (CAIDA) launched a series of *mapping projects*, aimed at exploring the Internet and drawing its global layout. Compaq's project was called *Mercator*, in honour of the geographer that drew in the 16th century one of the the most important maps of the world, including the recently discovered America: the Internet was acknowledged to be a 'new world' to be explored. Thanks to these and other projects, maps of the Internet are available, and its growth is now monitored.

The dynamics of the Internet have not stopped: at this very moment, routers, computers, cables, and satellite connections are constantly being added or removed, with the net effect of continuous growth. Despite the absence of a plan, this is not a completely random process. On the contrary, the Internet is a highly ordered and efficient structure. This emerging order must be the result of some regularity in the behaviour of individual agents that build the Net. There must be some small-scale mechanism that, iterated through a great number of interactions, ends up generating a structure that is organized at the large-scale level. Disentangling the general principles underlying those processes that shape network structures is essential to understanding their self-organization.

Even networks seeming to be completely static are subject in fact to some kind of dynamic process. The networks of genes, proteins, and metabolites in the cell, those of neurons in the brain, and those of species in the ecosystem, seem to be fixed in time: genetic regulation, metabolic pathways, neuron connections, or prey–predator relations are relatively stable. However, the networks within the cell live an explosive growth during the development of each individual and are constantly changing as the organism ages, and in reaction to the environment. The plasticity of the brain may decrease throughout life, but it is never completely lost. Extinctions or invasions of new species radically reshape foodwebs. Moreover, all biological nets change in the long run, for the action of natural evolution. Something similar happens in other apparently static networks, such as power grids or language networks. After power grids are set up, they slowly modify their shape due to accidents and technological evolution. The network of words changes as speakers change, and as the overall language evolves, introducing neologisms and new semantic relations.

In other networks, the change is mostly concentrated in the connections, while the set of vertices is almost immutable. For example, every day the banks of a given country set up a different pattern of money lending, in the interbank network. Occasionally, the set of vertices of this network can change, for example if a bank goes bankrupt or if a new one appears on the market. However, this change happens on a much longer timescale than the transactions (i.e. the rearrangement of edges): on a given day, the changes in the networks are mainly associated to the edges. Something similar happens in the networks of price correlations between stocks (correlation varies much more frequently than the actual set of stocks), in the World Trade Web (changes in economic relations between countries of the world are more common than the creation of new countries, through separation or federation), and in the airport network (in any one year, only a few new airports open, while a great number of flight connections change).

Exactly the opposite happens in another group of networks: in these graphs, new nodes are constantly added, and this process is much more relevant than the rearrangement of links. The most exact realization of this case is the network of citations within scientific papers. New papers appear every day, with citations to older ones, and once they are published, the respective citations cannot be changed any more.

Finally, there are networks where the dynamics of adding (or eliminating) nodes and connections happen at the same rate, giving place to a quite complicated process. The Web is constantly updated with creation and deletion of both new web pages and new hyperlinks. On some specific websites, such as Wikipedia, new articles and new connections between articles are created every day.

The range of possible network dynamics is extremely wide: network scientists have made measures, mathematical models, and computer simulations in order to grasp the basic mechanisms underlying this process, in the hope of understanding the principles of self-organization of these networks.

The rich get richer

At the beginning of the sixties, sociologist Harriet Zuckerman interviewed a number of scientists who had received the Nobel Prize. She aimed at finding what was so special in their way of working, and what the secret of their success in research was. She found a recurring theme in the answers of the Nobelists. A laureate in physics said: 'The world is peculiar in this matter of how it gives credit. It tends to give the credit to [already] famous people.' A laureate in chemistry added: 'When people see my name on a paper, they are apt to remember *it*, and not to remember the other names.' And a laureate in physiology and medicine specified: '[When you read a scientific paper] you usually notice the name that you're familiar with. Even if it's last it will be the one that sticks [...] You remember that, rather than the long list of authors.' 'The man who's best known gets more credit, an inordinate amount of credit,' concluded a laureate in physics.

Those observations led another sociologist (and later husband of Zuckerman), Robert Merton, to formulate in 1968 a brilliant law. Merton put forward that science is under the effect of a social mechanism dubbed the *Matthew effect*. The name comes from the Gospel of Matthew:

> For to all those who have, more will be given, and they will have in abundance; but from those who have nothing, even what they have will be taken away. (Mt, 25:29)

According to Merton, this mechanism plays a role in the distribution of prizes, financing, visibility, prestige... Scientists with a good number of these assets acquire even more of them with ease. On the other hand, those without them have a hard time acquiring and retaining them.

In 1976, physicist and science historian Derek de Solla Price found quantitative evidence that supported this view. Price analysed a large set of scientific papers, linked by mutual citations. He found that papers with a good number of citations at a particular time tended to acquire even more citations later, while those that had only a few at the beginning did not increase their citations that much. Price showed mathematically that this simple rule could explain the appearance of highly cited papers (hubs of the citation network). More precisely, he showed that this mechanism could explain why the distribution of the number of citations per paper displayed the characteristic fat tail of a power law.

Price's model was a variant of mathematical models developed earlier by statistician George Udny Yule and social scientist Herbert A. Simon, but it was not until 1999 that it became clear that this mechanism could explain the appearance of hubs, heterogeneity, scale-invariance, in synthesis of fat-tailed distributions, in a set of different networks. This realization is due to physicists Albert-László Barabási and Réka Albert. Barabási and Albert put forward a mathematical model of the growth of a network. They imagined a graph that starts with a small set of vertices (even just two or three), connected at random. New nodes are added at a steady rate to this initial nucleus, each of them carrying a given number of links. A simple rule establishes how new nodes are linked: incoming vertices prefer old ones that already have many links. This mechanism is called *preferential attachment* (Figure 10). In principle, new vertices can attach to any of the old ones, but the higher the degree of an old node, the higher the probability of attracting a new one. Occasionally, less connected nodes will receive new links, but most of the time hubs will be much more attractive.

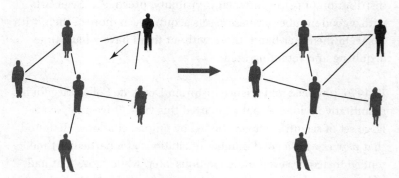

10. In the preferential attachment mechanism for network growth, new nodes connect preferentially with old nodes that have a high degree

This process can be studied mathematically and simulated on computer, and one can take measurements in real networks to see whether it is at work or not. At the beginning, all the nodes have more or less the same degree. However, during growth, some of them start to accumulate more links than others. The more connections a node has at a certain moment, the more it is capable of attracting new connections. This is why the preferential attachment is also called the *rich-get-richer* mechanism. As a consequence, the initially small differences in connectivity are progressively amplified. Thus, a hierarchy of different nodes emerges, with a large variety in their degrees, ranging from the least connected ones to those that have accumulated many links, i.e. the hubs. The resulting network is heterogeneous, with a power-law degree distribution.

The Barabási–Albert model shows that a bottom-up mechanism of growth can generate heterogeneity, without imposing any top-down blueprint. The global scale-invariance of the network is the outcome of the iteration of an individual, local choice: preferring more connected nodes to less connected ones. The model uses probability to allow for individual deviations from this behaviour: some nodes can decide to connect to low-degree nodes. However, the general tendency sets the outcome. As a further confirmation, one can check that the heterogeneity of the network disappears if it is grown without the preferential attachment rule. Indeed, new nodes connect to old ones at random, in such a way that the degree of old nodes does not influence their capacity to attract new links; the outcome is a homogeneous network in which every node ends up having more or less the same degree.

A widespread mechanism

Preferential attachment is the network version of a mechanism at work in many natural and social phenomena whose future evolution depends on their history. For example, the size of cities changes in time according to their present size: large cities experience large increases, small cities have small changes. Tomorrow's stock prices are also proportional, on average, to today's prices. This mechanism is also called *multiplicative noise*. There are various reasons why such a process may be at work in many networks. In some situations, having many links is the main way to be discovered by new nodes. A website linked to by many others is more easily found by people browsing the Web than is a less linked one. The same happens to highly cited papers. This increased visibility makes it easier to receive even more links or citations.

In other cases, links are attractive for their own sake. Sociologists have found evidence of *indirect mate choice* in which people choose their partner not only on the basis of that individual's personal features, but taking into account as well other people's opinion. For example, a study found that college-age women tend to rate higher a man in a picture where he is surrounded by many other women than the same man photographed alone. A long list of previous partnerships may put a person in a good position to attract even more partners. This is why preferential attachment is also called the *popularity is attractive* principle.

A more conscious reason for preferential attachment is that linking to hubs grants easier access to many other nodes. Since deregulation, in 1978, many airlines have adopted a *hub policy*, consisting in choosing well connected airports as their favourite destinations. The incentive is clear: since these airports provide access to a large range of destinations, connecting to them attracts more potential clients. Something similar happens in the web of interlocking directorates: a director that sits on many boards has access to a lot of information and a broad vision, something that makes it very attractive to hire him or her onto even more boards. Access is essential in the Internet, too. A large part of the Internet is set up and maintained by private companies, called *Internet Service Providers* (ISPs). When one of them sets up a new infrastructure, its priority is granting users rapid access to the information stored in the network. With this idea in mind, ISPs do not choose at random the routers to which they want to connect, nor do they just choose the closer ones. On the contrary, they select nodes that grant access to the greatest possible number of servers, in the smallest possible number of steps. And what is better than a hub to achieve this? In the case of the Internet, the data provided by the mapping projects seem to agree with the preferential attachment hypothesis. Figures show that vertices with many connections in a map issued at a certain time tend to have even more connections in the next map.

In some cases, a preferential attachment process comes in the guise of other mechanisms. Imagine someone wants to create a personal web page. A common strategy is to look at friends' pages and pick a nice one to use as a template. Since a person usually shares common interests with his or her friends, the new web page will likely keep most of the links of the template, maybe changing a few of them. So in the end, the new page will be a copy of the template, with a few changes. Now, this mechanism disguises a form of preferential attachment. In fact, to what pages does the template point to? Most likely to hubs. Only because hubs exist, and capture a large fraction of the connectivity of the WWW, any given page will point more likely to hubs than to weakly connected pages. So, if a page is duplicated, each copy brings even more links to hubs, resulting in an effective rich-get-richer process, so that the preferential attachment rule is recovered. The copying mechanism seems weird, but in fact it is a leading factor in several situations. For example, scientific citations included in an article are often drawn from other articles in the same field, and tend to consolidate the authorities of the sector.

But one of the most interesting applications is in genetic networks. Genomes often evolve through the process of *duplication and diversification*. During cell replications, all the DNA is copied to the new cells, but sometimes a mistake happens: a full gene of the original DNA chain is duplicated, and appears twice in the genome of the daughter cell (*duplication*). Most of the time, this new gene just produces redundant proteins that do the same as its twin. However, in further replications, one of the two can suffer a mutation that may allow its protein to perform new biological functions, for example interacting with different proteins than its former twin (*diversification*). This evolutionary mechanism has been observed in many cases. Now, its translation in terms of the protein interaction network is totally akin to a copying mechanism: a new node enters the network (the copied and mutated protein), with some of the links of its ancestor (the proteins with which it originally interacted), and some new ones, due to the mutation. Highly connected proteins have a natural advantage in this mechanism: it is not that they are more likely to be duplicated, but they are more likely to have a link to a duplicated protein than the weakly connected ones, and therefore they are more likely to gain new links.

Although the role of gene duplication has been shown only for protein interaction networks, there is evidence in favour of preferential attachment (either direct or in disguise) as well in metabolic networks. An obvious consequence of this linking procedure is that usually hubs are between the older nodes in the network, because they have had the opportunity to profit from 'first mover advantage'. Now, metabolic hubs are precisely primitive molecules, possibly incorporated in the genome during the evolution of the first forms of life, remnants of the RNA world such as coenzyme A, NAD, and GTP, or elements of the most ancient metabolic pathways, such as glycolysis and the tricarboxylic acid cycle. In the context of the protein interaction networks, cross-genome comparisons have found that, on average, the evolutionarily older proteins have more links to other proteins than their younger counterparts.

Preferential attachment is not the only mechanism at work in networks, and not all heterogeneous networks come from this mechanism. However, the Barabási–Albert model gives an important take-home message. A simple, local behaviour, iterated through many interactions, can give rise to complex structures. This arises without any overall blueprint, and it appears even if we allow for a certain level of randomness in the behaviour, accounting for individual deviation with respect to general trends.

When fitness matters

When an occasional sexual relation is at stake, people tend to be very tolerant with respect to certain features of the potential partner, such as political ideas, social class, or whether he or she smokes. However, these elements become very important when considering engagement or marriage. This is the message from an analysis of mid-nineties data by sociologist Edward O. Lauman. These data show that about three-quarters of the married couples in the US share a wide range of similar traits. These include belonging to the same social class, ethnic group, and education level, and even sharing similar levels of attractiveness, political ideas, and health behaviours, such as eating and smoking habits. On the other hand, when other types of sexual relationships are at stake, the proportion is much lower, although still high (more than half).

Homogamy, the tendency of like to marry like, is very strong even in societies where marriages are not combined and, theoretically, everybody could match with everybody else. In the turbulent college years, the prestige of an individual—as measured, for example, by the number of his or her previous sexual partnerships—can be a relevant driving force in shaping the sexual relations networks. But when people settle down, much stricter criteria come into force. While in the first case the rich-get-richer process can be at work, it can hardly explain the second case. Homogamy is a specific instance of *homophily*: this consists of a general trend of like to link to like, and is a powerful force in shaping social networks, according to much sociological evidence. In the Barabási–Albert model, the main criterion for linking to a node is the number of its links, but in many situations, other features, independent of the actual number of links, are much more important for attracting new links.

A consequence of the Barabási–Albert dynamics is that old nodes have a cumulative advantage over new ones. However, this is not always the case in practice. For example, old glories of the Web, such as Magellan or Excite search engines, are now mostly forgotten. Newer ones, such as Google or Yahoo!, have taken their place. Cumulative advantage can be completely overthrown when a new actor enters the game (think Facebook, for example). Newcomers often have some intrinsic feature that makes them much more attractive than older players. In this case, the connectivity of a network is not exclusively driven by the degree of the nodes, as in the Barabási–Albert model. On the contrary, a particular character of each node can play a very important role in its ability to gain links. This character is referred to as the *fitness* of the node, or as its *hidden variable*, a feature that shapes the structure of the network without being as evident as the number of links.

In 2002, physicists Guido Caldarelli, Andrea Capocci, Paolo De Los Rios, and Miguel Ángel Muñoz devised a model to generate a network only on the basis of its nodes' fitness. The basic recipe is identical to the random graph: within a set of nodes, all the possible pairings are considered, and a link is drawn or not between each pairing, according to a given probability. However, in this case, the probability is not fixed, but changes depending on the fitness values of the nodes in the pair. The first step in the model is distributing fitness to the nodes. This hidden variable may represent, for example, the income of an individual. If this is the case, it would be distributed between the nodes in such a way as to mimic the wealth distribution of a country: a few very rich nodes, then a certain number of upper-middle-class nodes, then lower-middle-class, and so on. The second step is defining the probability of linking. In order to simulate a highly segregated society, one could establish the following rule: the probability that two individuals form a social relation depends on the difference between their incomes; specifically, it is inversely proportional to it. That is, the higher the difference of income, the lower the probability they get connected. There is always some chance that two individuals with very different incomes could get connected, but in any case this will not be the leading trend: in general, homophily will triumph.

The fitness model may seem an overly simplified mechanism, but in some cases it works perfectly: for example, in the world trade web. In this case, the fitness is the GDP of the countries of the world. The linking rule is the following: the higher the fitness of two countries, the higher the probability that they get connected: this would be a kind of *fit-get-richer* mechanism, in which the countries with highest GDPs tend to build more commercial links with each other. This is a different mechanism from homophily, because while high-GDP countries establish many links with similar ones, low-GDP countries do not link with other low-GDP countries; they just stay poorly linked. Physicists Diego Garlaschelli and Mariella Loffredo showed in 2004 that these ingredients are enough to predict with great precision the features of the world trade web, provided that one feeds into the model the overall number of existing commercial links at any given time. For example, the model predicts exactly the shape of the degree distribution of the network. This suggests that the basic mechanisms underlying the self-organization of the world trade web are captured by the simple dynamics of the fitness model.

Like preferential attachment, it is unlikely that the fitness mechanism is at work in all real-world networks. While the Barabási–Albert model is plausible when applied to growing networks, the fitness model works also for static networks, where the number of nodes is mainly fixed. Nevertheless, they can be at work at the same time: in 2001, physicists Ginestra Bianconi and Albert-László Barabási introduced the idea of fitness in a preferential attachment model, showing that a mixture of the two effects gives a good prediction of the Internet's topological properties. It has to be noticed that the fitness model does not always yield a power-law degree distribution. A broad range of distributions of fitness and linking rules generates it, but many others do not. However, this is not a limitation, but rather a positive aspect of the model, that allows it to be applied also to non-heterogeneous networks such as the world trade web.

A variety of strategies

The myth of the girl (or boy) next door has passed into history. In the mid eighties, the number of people that reported meeting their spouses in their own neighbourhood was already negligible, about 3 per cent in France, according to a 1989 study by sociologists Michel Bozon and François Heran. However, just three decades earlier, this case was quite common: Bozon and Heran found proportions of 15 to 20 per cent between 1914 and 1960. In a number of countries in the world, this is still the situation. In some settings, marriages (and social relations in general) are driven neither by some kind of popularity nor by similarity criteria: when geographical constraints matter (for example, if regular access to long-range transport is not available), one is forced to make friends with neighbours and fellow villagers.

In these cases, the vertices of a network are embedded in physical space and this has many important consequences. In some cases, one can connect at almost no cost with anybody else (making friends on a virtual social network). But in other situations, a long-distance connection is quite costly. The properties of the networks in the two cases are then quite different. Many infrastructure networks (trains, gas tubes, highways, etc.) display this bias, since they are embedded in the physical space. Other networks are embedded in time: for example, scientific papers are published on a certain day, and this creates a bias in linking, namely that new papers can cite only older ones, while old papers cannot cite newer ones.

Other biases and strategies can influence network formation. Sociologists have identified two basic incentives for linking in social networks: *opportunity-based antecedents*, that is the likelihood that two people will come into contact, and *benefit-based antecedents*, that is some kind of utility maximization, or discomfort minimization, that leads to tie formation. The global *optimization* of some quantity can play an important role in shaping technological networks: for example, the pressure to minimize search costs on the WWW can lead to a tendency to optimize the shortest path lengths and the link density.

Finally, it may even be that apparent self-organization comes from complete randomness. Imagine that a company releases a new social network and provides nicknames to 100,000 people. Then the company gives permission to one individual to connect with 1,000 others, permission to two individuals to connect with 500, permission to three to connect with 333, permission to four to connect with 250, and so on. People do not know who is behind the nicknames, so they choose their partners at random. Obviously, there is no self-organization in this process: the rules established by the company determine the structure. Still, the final network has *by construction* a power-law degree distribution. This example shows that, in some cases, power laws do not necessarily mean self-organization.

All this large range of strategies, biases, processes, and motivations must be taken into account when trying to understand the features of a network by modelling its behaviour. It may even be that each individual network needs a model of its own. However, it is likely that some very general mechanisms, such as preferential attachment or fitness-related dynamics, play a role in the formation of large classes of apparently unrelated networks. The models described in this chapter are simple explanations of the ways in which local mechanisms, without global planning, can indeed generate large-scale, complex, ordered, and efficient structures.

Chapter 7
Digging deeper into networks

Who are your friends?

For every white American with a sexually transmitted disease, there are up to twenty African-Americans with the same condition in certain areas of the US, according to several studies carried out in the nineties. This figure is the outcome of persisting racial inequalities. However, the actual mechanisms of contagion that generate such a big difference are still partially obscure. In 1999, sociologists Edward O. Laumann and Yoosik Youm found an interesting piece of evidence: less sexually active African-Americans (those who had only one partner in the previous year) were five times more likely to have intercourse with more sexually active African-Americans (those who had four or more partners in the previous year) than whites in the same situation. In other words, in the sexual interaction network of the whites, the *periphery* of less active people was partially separated from the *core* of active individuals. On the contrary, these two groups were more connected in the African-American network. The reason for this difference is unclear, but its consequence is straightforward: in the first network, sexually transmitted diseases thrive mainly within the core, while in the African-American one, they also spill over to the periphery.

This is a case in which the degree of the nodes in the networks is not the most relevant quantity in understanding the situation. Individuals with exactly the same number of sexual connections have a different exposure to infection depending on whether they are African-Americans or whites. In situations like this, it is not enough to know how many 'friends' you have (your degree): it is necessary also to know how many friends your friends have. The degree distribution provides a great deal of information about the overall structure of a graph, for example whether it has hubs or not. However, it does not tell everything about that graph. For example, consider two networks with the same number of nodes and edges: the nodes may have exactly the same degrees, but the edges can be arranged in such a way that the outcomes are very different graphs. The degree is a local feature of a vertex. In order to capture the more subtle structure of networks, one has to dig deeper, and find measures that describe the surroundings of a node: its nearest neighbours, the neighbours of its neighbours, etc.

In the whites' sexual network, low-degree nodes tend to connect with low-degree nodes, and high-degree nodes with high-degree nodes. This phenomenon is called *assortative mixing*: it is a special form of homophily, in which nodes tend to connect with others that are similar to them in the number of connections. By contrast, in the African-Americans' sexual network, high- and low-degree nodes are more connected to each other. This is called *disassortative mixing*. Both cases display a form of correlation in the degrees of neighbouring nodes. When the degrees of neighbours are positively correlated, then the mixing is assortative; when negatively, it is disassortative.

Usually, the presence of these patterns of mixing is the outcome of some non-trivial mechanism acting in the network, possibly a form of self-organization. In random graphs, the neighbours of a given node are chosen completely at random: as a result, there is no clear correlation between the degrees of neighbouring nodes (although the finite size of a graph can disguise this to some extent). On the contrary, correlations are present in most real-world networks. Although there is no general rule, most natural and technological networks tend to be disassortative, while social networks tend to be assortative. For example, highly connected web pages, autonomous systems, species, or metabolites tend to be linked with less connected nodes of their networks. On the other hand, company directors, movie actors, and authors of scientific papers tend to link with those similar to them in connectivity: the higher the degree of an individual, the higher that of his or her neighbours in the network.

Degree assortativity and disassortativity are just an example of the broad range of possible correlations that bias how nodes tie to each other. For example, Laumann and Youm also showed that African-Americans, much more than other groups, tend to have partners from their own community. As a consequence, when an infection enters the community, it gets 'trapped' in it. This simple effect alone makes the likelihood of African-Americans having a sexually transmitted infection 1.3 times greater than the figures for white Americans. In this case, the correlation does not arise from the degree, but is a form of homophily with respect to an intrinsic character of each node, namely its ethnic identification. Another example is correlation with respect to body mass: it has been shown that people with similar body mass index tend to establish social bonds with each other more frequently than with other people. One must note that correlations do not always need to be positive, favouring homophily: for example, in foodwebs, edges connect plants to herbivores, and herbivores to carnivores, but very few connect herbivores with herbivores, or plants with plants.

Who are your friends' friends?

Cosimo de' Medici, the man who led his family to take over Florence in 15th century, was described as an 'indecipherable sphinx'. Although he rarely spoke publicly, and never committed openly to almost any form of action, he was able to build around him a strong party that made him the *pater patriae* (father of the nation) of the most important city of the Renaissance. In 1993, social scientists John F. Padgett and Christopher K. Ansell analysed the information about marriages, economic relations, and patronage links that connected the Medici to the other powerful families of the city. They found that Cosimo's family was at the centre of a network of ties with many of the leading lineages. More importantly, without the Medici connection, most of the time those families were weakly related, or even opposed to each other. Cosimo's reserved attitude helped him to establish relations of alliance and control with everyone.

The network with the Medici at the centre is an example of an *ego network*, a graph composed of a set of nodes with direct ties to a central one (the *ego*), as well as ties linking them to each other. Whenever one of the latter ties is missing (i.e. two neighbours of the ego are not neighbours to each other) the network has a *structural hole*. Cosimo's network was full of structural holes, and his family was able to use them to implement a *divide et impera* (divide and reign) strategy: the Medici were seen as a third party in many conflicts, and families had to ask their mediation in their relations with other families.

However, being surrounded by many structural holes is not always beneficial for an individual. Adolescent girls whose friends are not friends with each other are twice as likely to commit suicide, according to a 2004 study. A possible explanation of this finding is the exposure to conflicting inputs from unrelated friends. Another example comes from workers' unions: when workers' networks lack structural holes (that is, *egos* are surrounded by nodes with abundant mutual ties), then a powerful, well coordinated, and communicative organization arises. In general, different patterns of structural holes point to different situations. For example, a scientist working in a specialized field is usually connected to other scientists in the field, which are likely connected to each other. On the other hand, a scientist working in a highly interdisciplinary field is probably connected to experts of various areas not necessarily in contact with each other.

In all these cases, it doesn't matter how many friends you have (your degree), or who they are (for example, whether their degree is similar to or different from yours). What really matters is who your friends' friends are; in particular whether or not your friends are also friends to each other. This concept is often referred to as *transitivity*, or *clustering*. Let us consider an individual with two friends: they form a *connected triple*. If the two friends are friends with each other, then they also form a *transitive triple*, or *triangle*. The quantity of triangles in a network compared with the overall number of connected triples is the basic ingredient of the *clustering coefficient* of that network: this is a measure of the density of triangles in that graph, its overall transitivity. In random networks, the connections between the nearest neighbours of a node are as random as those between any two nodes. As a consequence, these graphs have just the quantity of triangles that emerge from a purely random disposal of edges. On the other hand, the clustering coefficient of almost all real-world networks is higher than their random counterparts. This suggests that some non-trivial process, possibly a form of self-organization, is at work in generating this extra transitivity.

The high clustering of many networks suggests the presence of groups where 'everybody is friends with everybody else'. At first sight, this picture seems to contradict the small-world property of networks: are networks 'open' worlds in which everybody is a few steps from everybody else, or are they the sum of tightly knit, segregated groups? In reality, there is no real contradiction between these two features: one can see this by having a closer look at the Watts–Strogatz model. The model starts by considering a circle of nodes, each of them connected with its first- and second-nearest neighbours, such as remote villages interchanging goods with their neighbours. This is a fully clustered structure, in which all the commercial partners of one village are commercial partners with each other. The model then allows for rewiring a few links to randomly chosen nodes: a few villages open paths to other faraway villages and bring goods there, declining to do business with one of their neighbours. A few paths are enough to reduce abruptly the distance between any two villages, but on the other hand we can think that the local tight commercial structure is disrupted, that is, its clustering goes down. However, Watts and Strogatz found that the decrease in the clustering is less pronounced than the decrease in the average distance. In practice, in order to make a noticeable drop in the transitivity, one has to rewire almost all the nodes. At this point, only random connections are present in the network. Since it is a random graph, we do not expect a large clustering. The take-home message is that networks (neither ordered lattices nor random graphs), can have both large clustering and small average distance at the same time.

Another interesting point about clustering is that in almost all networks, the clustering of a node depends on the degree of that node. Often, the larger the degree, the smaller the clustering coefficient. Small-degree nodes tend to belong to well-interconnected local communities. Similarly, hubs connect with many nodes that are not directly interconnected. In the Internet, for example, low-degree autonomous systems usually belong to highly clustered regional networks, interconnected by national backbones. A similar structure is likely to be present in many networks, where clustering decreases with increasing degree.

Who are your friends' friends' friends'...?

Money does bring some happiness, but being surrounded by happy people gives much more of it. Earning US$5,000 more per year increases the chance of being happy by just 2 per cent, according to a 1984 estimate, while having a happy friend increases it by 15 per cent, according to a 2008 study by sociologists Nicholas Christakis and James Fowler. The two scientists asked more than 12,000 people from Framingham, Massachusetts about their subjective feeling of happiness. Moreover, they mapped who was a friend, spouse or sibling of whom. By drawing this network, the two found that connected people tend to have similar feelings: happy people tend to group together, as, on the other hand, unhappy people do. Christakis and Fowler found even more interesting evidence. The happiness of an individual is influenced by the happiness of people that are not their immediate neighbours. The 'happiness effect' two steps away (friends of friends) is about 10 per cent; three steps away (friends of friends of friends), it is about 6 per cent. The effect fades only at the fourth step. The two sociologists and other scientists found similar results for obesity, smoking habits, and word-of-mouth advice (such as finding a good piano teacher or finding a home for a pet): in all these cases, influence and information arrived to an individual from three degrees away. This *three degrees rule*, found in several social processes, is an example of *hyperdiadic spread*: that is, the diffusion of a phenomenon beyond *dyadic* relations, those that connect nearest neighbours. In this case, it is neither the degree of a node, nor the degree of its neighbours, nor the connections between its neighbours that matter. The influence goes beyond the immediate circle of each node. Actually in many phenomena, it goes even beyond the third degree. For example, a highly infectious disease can form longer chains of contagion; similarly, the spread of nutrients in a foodweb can span all the network.

In this kind of dynamics, a node can be more or less important depending on the number of chains passing through it. In order to capture this idea, sociologist Linton C. Freeman introduced the concept of *betweenness centrality* of a node. One takes all the pairs of nodes in a network and counts the shortest paths connecting them. The betweenness centrality of a node is basically the proportion of shortest paths that cross that node. The higher this proportion, the more central is the node. According to this measure, the Medici are the most central family in the network of lineages in 15th-century Florence. In this case, the betweenness centrality is a measure of the potential to slow down the flow or to distort what is passed along, in such a way as to serve the node's interests. Several studies show that the centrality of a firm in economic networks predicts well its ability to innovate (as measured by number of patents secured) as well as its financial performance. Interestingly, between 1980 and 2005, East Asian countries experienced huge increases in their centrality in the world trade web, while the centrality of most Latin American countries went down. However, the trade statistics of these two regions displayed similar patterns: the big difference in their development was not tracked well by macroeconomics statistics, while the network-based approach captured it. Moscow became in the Middle Ages the most central node of the river transportation network of central Russia, according to a 1965 study. Very probably this set the scene for the future importance of the city.

Central nodes usually act as bridges or bottlenecks: they are almost compulsory stops in the traffic on a network. For this reason, centrality is an estimate of the load handled by a node of a network, assuming that most of the traffic passes through the shortest paths (this is not always the case, but it is a good approximation). For the same reason, damaging central nodes (for example, extinguishing a central species or destroying a central router) can impair radically the flow of a network. Depending on the process one wants to study, other definitions of centrality can be introduced. For example, *closeness centrality* computes the distance of a node to all others, and *reach centrality* factors in the portion of all nodes that can be reached in one step, two steps, three steps, and so on. More complicated definitions are also available.

The behaviour of centrality in many real-world networks is a further signature of their heterogeneity. In many real-world networks it exhibits the characteristic long tail of a heterogeneous distribution. The average centrality is not a valid estimate for the centrality of any node, because this magnitude varies a lot around the average: a few nodes are the main bottlenecks of almost all the shortest paths in the network, and a full hierarchy of less central nodes goes down from them. Given the importance of the more central nodes, it is natural to ask whether they are the same as the hubs of the networks. In many situations, this is in fact the case. For example, highly connected autonomous systems also act as bridges between regional networks; or polysemic words, with their many connections to other words, bring together separated areas of language. But this is not a general rule. A notable exception is the airport network: in this case, certain low-degree airports have exceptionally large betweenness. In 2000, the airport with highest centrality was Paris, a hub connected to more than 250 other cities. But the next highest was remote Anchorage, in Alaska, a medium-sized airport with just 40 connections. Other airports similar to it appear in the list of the most central ones. How is this anomaly explained? Alaska has many airports for internal flights, but Anchorage is its only bridge to the rest of the US, so many paths cross this airport. The anomaly is the result of the existence of regions with a high density of airports but few connections to the outside world.

What group(s) do you belong to?

In 1972, two karate instructors in a university club in the US were so much in conflict that they decided to split their club into two different ones. This event, pretty unexciting for most of the world, was a goldmine in the eyes of social scientist Wayne W. Zachary. In 1977, he published a pioneering study in which he gave an unconventional view of this event.

In 1970, the karate instructor, Mr Hi, asked the president of the club, John A., to raise the prices of lessons in order to provide a better salary. All he received was a denial. As time passed, the entire club became divided over this issue, and after two years the supporters of Mr Hi formed a new organization under his leadership. During all that time, Zachary collected information about the karate lessons, the meetings, the parties, and the banquets of club members, and identified as good friends those that also met outside the club. At this point, he was able to draw a precise network of the friendships within the club: the resulting structure was clearly divided into two groups built around the two instructors, each of them composed of people who were friends with each other and with one of the instructors, with few connections with people in the other group (Figure 11). When the club split in two, the people divided almost exactly along the lines that separated the two groups.

Zachary's method was able to predict the club division almost perfectly, on the basis of the structure of the network alone. Since then, researchers have been striving to find a general method to identify *communities* or *modules* in networks. In Zachary's case, they could be seen just by examining the map, but other instances are much more complicated, and a general solution is still to be

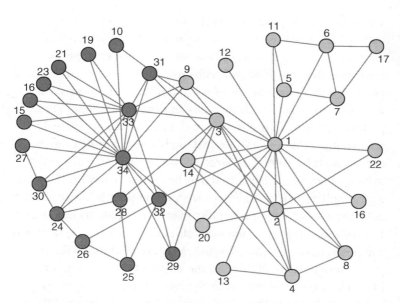

11. The structure of friendships of a karate club studied by anthropologist Wayne Zachary allows us to predict the separation of the group into two communities

found. All real-world networks display some level of modularity. Alaska is obviously a specific module within the airport network structure, such as other regions well connected to the inside but not to the outside. Foodwebs are divided into *compartments*: groups of species interacting more strongly among themselves than with others. Social networks are divided into *cliques*: for example, studies with adolescents have showed that their behaviours are strongly influenced by the modules to which they belong. The neuronal network is divided into big areas, often corresponding to specific functions. The genetic regulatory network is divided into subnetworks, associated to specific functions or diseases. Degree, correlations, clustering, and centrality provide information on single nodes, their immediate surrounding, and their position with respect to the overall network, but they do not capture the discrete structures into which the overall graph is divided.

The simpler form of module is the *motif*, a pattern of connections within a few nodes repeated throughout the network. In foodwebs we frequently find a diamond-like structure: for example, a carnivore eats two different herbivores, and they both eat the same plant. Another common motif is a simple chain of three species: a big fish eats a small fish that eats an even smaller one. These patterns are not the result of pure chance: motifs appear with a much higher frequency in a real foodweb than in its random counterpart. In general, in large networks, one can isolate many subsets of nodes and edges that may be candidate motifs. However, a given subgraph is considered a relevant motif only if it occurs in a network with a higher probability than in its random counterpart. In the Web, a very common example is the *bipartite clique*: this is composed of two groups of websites, where all those in the first group have links to all those in the second. Often, this motif identifies a group of 'fans' with the same interests (say, blogs on rafting), pointing to their 'idols' (say, websites of rafting magazines). Networks that regulate genes are almost completely built out of motifs. When the bacterium *E. coli* is in a stressful condition, a specific genetic circuit senses the stress and coordinates the production of certain proteins. These proteins coalesce to build flagella, a kind of moving tail that allows the bacterium to swim away in search of better conditions. This same genetic circuit, the *coherent feed-forward loop*, is present in many other bacteria and several other organisms. Evolution seems to have selected specific motifs, because of their optimal properties, (e.g. because they use the smaller number of genes necessary to perform a certain function). Moreover, a clear advantage of modularity is that motifs can combine to give rise to new functions, and damage to one of them is not propagated to the others.

Motifs are a kind of small-scale, local, repeated modules. But when people think about communities, they usually aim at finding great partitions of a network, such as compartments of a foodweb, online communities, disciplinary areas, etc. These structures do not show a regular, repetitive pattern. The task of finding them is easier if we have some clue, for example if the members of the community are self-identified by some element. This might be a widget all the members of the community add to their blogs, a common way of dressing, etc. However, most of the time this information is neither available nor explicit, and we have to dig into the network structure to find the modules. The general objective of community identification is to find sets of nodes that are more highly interconnected between themselves than with each other, such as in the karate club network. This verbal formulation is easy, but the mathematical translation of the concept is elusive, such that a definitive community detection method has not yet been found. Some methods aggregate nodes to fulfill an optimality criterion. Others split the network into groups, and then split the groups further, and then split them even more, to create a genealogic tree of nested communities for the partition. Others place imaginary springs between nodes and look at the clusters formed after the relaxation of the system. In general many other options are available. An interesting technique, that cleverly uses the topology of the network, is based on computing the *edge betweenness*. That is, finding the edges through which most of the shortest paths pass. The links with highest edge betweenness are akin to the weak links connecting otherwise separated groups in Granovetter's work. If one cuts a few edges with high betweenness, then the network splits into a certain number of isolated clusters: these are suitable candidate communities. One can continue cutting the higher-betweenness edges to find more detailed structures nested within the larger ones.

An interesting application of community finding is the analysis of the US political blogosphere. Physicist Lada Adamic found a clear separation between Democrats' and Republicans' blogs. The resulting structure showed two large groups with very few connections to each other. Moreover, the structure related to liberals' blogs was found to be less cohesive than the conservatives' one. For instance, in the part of this blogosphere dedicated to abortion, pro-life blogs show a denser interconnection than pro-choice blogs. As a result, an online campaign is likely to spread more easily in the first group than in the second. Another study analysed communities of students in US schools, to see whether ethnicity shaped social networks. In both very diverse and very homogeneous schools, ethnicity seemed to be irrelevant. On the contrary, segregation was visible for intermediate values of diversity. In the metabolic network, communities have been found to correspond to specific functions (carbohydrate metabolism, nucleotide and nucleic acid metabolism, protein, peptide, and amino acid metabolism, lipid metabolism, aromatic compound metabolism, monocarbon compound metabolism, and coenzyme metabolism). Finally, company stocks have been clustered on the basis of price correlations, finding modules corresponding to the various business areas, such as banking, mining, distribution, financial, etc.

The definition of community as a subgraph 'more densely connected internally than externally' is very general, but fails to capture some special modules. Think of a telephone chain between friends, in which the first calls the second, the second calls the third, and so on: such a chain would most probably not be classified as a community, according to this definition. Another example is the web pages of competitors working in the same business: obviously, they do not have incentives to link to each other, although they clearly belong to the same community. In addition, real-world communities are much more complex than dense clusters. Several partitions are possible at the same time; nationality, social class, gender, job, political ideas: all could classify the same set of people. Moreover, communities can overlap: an individual can belong to multiple nationalities or affiliations at the same time. Finally, nested communities are possible: for example, regional identities within national identities.

Albeit a strong simplification, the graph representation is still capable of capturing many relevant features of a system. A close look at a graph provides plenty of information and more details arise when more complex measurements are performed. Almost all the time, real-world networks deviate from their random counterparts, suggesting the existence of some kind of built-in order. Again, all these networks are not blueprinted: these deviations very likely arise from self-organization processes. Finding new regularities in graph structures and revealing the underlying mechanisms are some of the ongoing challenges of network science.

Chapter 8
Perfect storms in networks

Settings for surprise

The island of Barro Colorado is a piece of rainforest in the middle of the waters of the Panama Canal. When a nearby river was dammed, just a few hilltops remained uncovered. The island has become an open-air experiment about what happens to a forest when it is fragmented into small pieces, as when highways, buildings, fields or mines substitute the original vegetation. A few years after the inundation around Barro Colorado, the population jaguars and pumas had shrunk dramatically. As a consequence, their prey thrived: now, in the island there is plenty of specimens of a large rodent, called agouti. These animals love the big seeds of acacias, so their boom is a big problem for acacias to successfully reproduce, as well as for microorganisms colonizing their seeds. As the acacia population shrinks, plants producing smaller seeds occupy their place, and animals eating them also increase in number. The original alteration of the ecosystem extends in all directions in the foodweb of the island.

Domino effects are not uncommon in foodwebs. Networks in general provide the backdrop for large-scale, sudden, and surprising dynamics. Pathogens spreading in transport networks, blackouts in power grids, large conflicts, or unexpected cooperative efforts in social systems: networks seem to be the ideal setting for 'perfect storms'. Network nodes can represent individual entities (people, computers, species, genes...) exchanging material or information (information packets, energy, etc.), or they can represent locations (countries, airports...) exchanging individual entities (goods, travellers...). Within this very broad classification, the range of possible dynamics is enormous. Why are networks the natural playground for all these dynamics? How does the graph structure influence these processes? A general answer is impossible, but in many cases we can see that the heterogeneous, non-random organization of the underlying network makes a big difference to all the phenomena taking place on top of it.

Failures and attacks

On 18 July 2001, a train derailed in an underground tunnel in Baltimore (US), and began a fire. Soon after, the Internet was slowed down in several states along the US east coast. The fire had burnt optic cables passing through the tunnel, that belonged to several of the most important Internet Service Providers of the country. As a consequence, the accident created a domino effect that spanned a large part of the US. The Internet is constantly exposed to similar accidents. A percentage of routers are always out of operation at any time, for a broad range of reasons. Potentially, each one of these accidents may be as serious as Batimore's derailment. Still, such macroscopic damages are rare. The network seems to tolerate a certain amount of chronic dysfunction without too many problems. It relies somehow on alternative paths, allowing traffic to get around failures. Still, like most networks, the Internet does not have many redundant links, and is not highly dense either. With these features in mind, it would be natural to expect it to break down easily.

While the Internet seems to be relatively resistant to errors and accidents, a carefully designed attack can wreak terrible damage. On 7 February 2000, an enormous number of users logged on to the Yahoo! website. There were so many that the company servers were not able to answer these requests and the web page went down. In the days that followed, a set of other web pages, ranging from eBay to CNN, went down for the same reason. After two months, the police discovered that the logons were artificial and came from a 15-year-old Canadian hacker, whose nickname was Mafia Boy. He did not need to burn any cable to block the Internet: what he did was enough to bring down the websites that attracted most of the traffic on the WWW.

As with the Internet and the WWW, most of the real-world networks show a double-edged kind of robustness. They are able to function normally even when a large fraction of the network is damaged, but suddenly certain small failures, or targeted attacks, bring them down completely. For example, genetic mutations arise naturally throughout life (and some of them can even delete certain proteins from the cell) or are produced artificially (as in the case of a genetic technique called *gene knockout*, that turns off the function of a whole gene in lab rats). Still, organisms display a great tolerance to many mutations, and to an unexpectedly large number of knockouts. Most of the time they continue to work normally, in overall terms. On the other hand, certain specific mutations are capable of completely disrupting the workings of a cell. The brain loses neurons all the time: a stressful experience for any given organ, such as getting drunk occasionally, can kill a considerable number of cells. But after the hangover everything works fine again, usually. In Parkinson's disease, a large portion of the neurons can disappear without the patient even noticing. But when this portion exceeds a certain threshold, then the disruptive condition starts to become manifest.

In this respect, networks are very different from engineered systems. In an airplane, damaging one element is enough to stop the whole machine. In order to make it more resilient, we have to use strategies such as duplicating certain pieces of the plane: this makes it almost 100 per cent safe. In contrast, networks, which are mostly not blueprinted, display a natural resilience to a broad range of errors, but when certain elements fail, they collapse. How many errors can a network tolerate without even noticing the problem? And what are the elements that cause the collapse? With the objective of answering these questions, scientists have simulated failures by removing nodes from network maps and observing what happens. After the removal of a fraction of nodes, they check whether the surviving nodes are still connected (that is, whether a giant connected component is still present in the network) and close (that is, whether the average distance is still small). In order to simulate errors, the nodes are removed at random. When this is done to a random network, after a few removals the distance increases quickly and the graph breaks down in many disconnected components. A random graph of the size of most real-world networks is destroyed after the removal of half of the nodes. On the other hand, when the same procedure is performed on a heterogeneous network (either a map of a real network or a scale-free model of a similar size), the giant connected component resists even after removing more than 80 per cent of the nodes, and the distance within it is practically the same as at the beginning. The scene is different when researchers simulate a targeted attack, as in the strategy of Mafia Boy. They started by removing first the most 'important' nodes (hubs) of the network. In this situation the collapse happens much faster in both networks. However, now the most vulnerable is the second: while in the homogeneous network it is necessary to remove about one-fifth of its more connected nodes to destroy it, in the heterogeneous one this happens after removing the first few hubs.

Highly connected nodes seem to play a crucial role, in both errors and attacks. They are the 'Achilles heel' of most heterogeneous networks exposed to targeted attacks. In these networks, hubs are mainly responsible for the overall cohesion of the graph, and removing a few of them is enough to destroy it. On the other hand, hubs are also the 'ace in the hole' of these networks, when they are exposed to errors and failures: when nodes are removed at random, most of the time the selected nodes come out from the large population with low degree, so as long as hubs are kept untouched, the network stays together. This behaviour becomes clearer considering that the degree is usually correlated with the betweenness. High-degree nodes are most of the time bridges through which many paths of the network pass. When random damage is applied to a network, it will rarely affect one of the few hubs. While hubs are unaffected, they provide the necessary connectivity: there is no need for many redundant connections; paths crossing hubs keep the working areas of the damaged network connected. In those few networks in which some low-degree nodes have high betweenness and act like bridges (as certain airports do), attacking hubs still causes serious damage, but the most lethal strategy is attacking the most central nodes.

Domino effects

The possibility of a sudden transition from a resilient behaviour to a global collapse should ring some alarm bells. In ecosystems, a certain rate of extinction is inevitable: one in each million of species becomes extinct every year, according to some estimates. Usually, foodwebs rearrange after these events, and most of the species do not suffer major damage from these natural extinctions. But large-scale collapses are possible too: about 250 million years ago, more than 90 per cent of the species disappeared in a relatively short period, the famous Permian extinction. Five massive extinctions of this kind have been registered in the last 500 million years. Researchers have argued that external factors may be the cause, such as the much-debated meteorite that may have made dinosaurs extinct. However, a network explanation is also possible. Cases of extinctions in chain, or *co-extinctions*, are not unknown to ecologists. For example, the introduction of the virus of myxomatosis to control the population of rabbits in England in the mid-20th century ended up making the big blue butterfly (*Maculinea arion*) extinct in 1979. The virus decimated rabbits, and as a consequence the tall grass they ate spread in the fields. This destroyed the habitat of ants, that used to make nests in low grass, where the sun could reach. Ants had a mutualistic relation with blue butterflies' larvae: they took care of the larvae, which responded by providing liquid food to the ants. The disruption of their habitat gradually impaired the reproduction of the butterflies, bringing them to extinction. This is not a coextinction in the literal sense, since rabbits did not disappear due to mixomatosis and blue butterflies have been partially reintroduced. However, it gives an idea of how far damage to foodwebs can go. A large-scale version of this story, with a chain of extinctions that depletes almost a full ecosystem of species, is a possible alternative explanation to the great extinctions of the past. This should also be taken into account when massive attacks on ecosystems are voluntarily carried out by humans, as in the case of too much fishing currently depleting marine ecosystems at an unprecedented scale.

Several other dynamic processes on networks could give rise to similar *cascading failures*, or *breakdown avalanches*. A typical example is a large-scale blackout: the failure of a power station overloads another one, which fails in its turn, propagating the overload throughout a large part of the network. In this phenomenon, the failure of a node results not only in loss of interconnection or reduction of the average distance but also in a domino effect. The *systemic failure* of economic networks experienced during financial crises is another instance of this phenomenon. The same can happen in *congestion phenomena*, such as cars collapsing certain points of the street network, people collapsing a subway station during a special event, or online traffic collapsing certain Internet services. In all these cases, studies have shown that hubs are crucial, both because they reduce transit times and because they are first in becoming saturated.

Epidemics

In 1347, one of the most devastating plagues in human history appeared in Constantinople. During the following three years, the Black Death moved to Europe, leaving a large fraction of its population dead. The disease covered Europe like a wave, at a velocity of 200–400 miles per year (Figure 12 left). This picture is much different from that of modern pandemics. The 1918 influenza that is estimated to have killed 3 per cent of the world population took just one year to spread, reaching even isolated Pacific islands in that time. The 1957 flu virus, also called 'Asian flu', swept the globe in about six months. More recent outbreaks, such as 2009's swine flu, leapt from one side of the planet to the other in a few weeks (Figure 12 right). While the Black Death travelled with pilgrims, merchants, and sailors, lurking in ships and carriages, at a few miles per day, modern diseases can rely on much more efficient means of transportation, such as highways, trains, and aeroplanes. In the 14th century, physical distance was a leading factor in the spread of an epidemic. In the modern networked world, an infection can jump on a plane and reach the opposite side of the planet in a few hours.

Epidemics spread in networks both at the global level (for example, through the airport network) and at the local level: infectious diseases that jump from person to person depend on individuals' social networks. For example, flu spreads partially through face-to-face contact between individuals, while HIV spreads in the network of unprotected sexual contacts. In 2001, the Cabilan physicists Romualdo Pastor Satorras and his Italian colleague Alessandro Vespignani decided to study the problem by modelling and simulating the spread of a disease in a social network. They introduced just the minimal ingredients of an infectious process: at the beginning, a few individuals of a social network get infected; if a healthy individual is in contact with one of them through a

12. The 14th-century bubonic plague swept Europe like a wave (left), while 2009 swine flu was more similar to a fire throwing sparks from one side of the planet to another (right): the difference is due to the dramatic change in human transportation networks

link, he or she has a certain probability of being infected; on the other hand, infected individuals have a certain probability of recovering. This model of infection is called *SIS*, because each individual passes through the cycle *susceptible–infected–susceptible* (a healthy individual is 'susceptible' to being infected). The process represents infections such as the common cold, from which people usually recover. It can be further complicated to represent other diseases, for example by introducing the possibility that people die or are immunized. However, the general direction of the results are not changed by these modifications.

Pastor Satorras and Vespignani found that, after an initial phase of expansion, the virus can either be eradicated—it shrinks and finally disappears from the population—or become endemic—it sustains itself and infects a certain fraction of the population indefinitely. The disease is said to be below the *epidemic threshold* when, for every infected individual, fewer than one person gets infected: in this case, it becomes extinct. The disease is above the epidemic threshold if every infected individual passes the disease to more than one individual: in this case, it thrives. If vaccines are available, the disease can be pushed below the threshold by means of campaigns that immunize a sufficient portion of the population. Very contagious diseases are usually the hardest cases, because they have a low epidemic threshold and so they become endemic very easily. If eradication is too hard, pushing the disease closer to the threshold still has a positive effect: that is, reducing the proportion of people indefinitely affected by the endemic disease.

In their study, Pastor Satorras and Vespignani found that the epidemic threshold depends crucially on the features of the underlying network. When the SIS dynamics are performed on a random network, a clear threshold is found that allows us to estimate how many individuals have to be immunized to extinguish the disease. But when the dynamics are performed on a heterogeneous network, then the threshold almost disappears: it is much lower than in the random graph; moreover, the larger the size of the system, the lower the threshold. In a large enough network, the threshold is so low that it is almost unavoidable to have a proportion of infected individuals. The disease cannot be pushed below such a low threshold without immunizing almost all the population. In epidemics, as in many other dynamics, heterogeneity makes a difference. Studies of errors and attacks have shown that hubs keep different parts of a network connected. This implies that they also act as bridges for spreading diseases. Their numerous ties put them in contact with both infected and healthy individuals: so hubs become easily infected, and they infect other nodes easily. The *super-spreaders* identified by epidemiologists are likely the hubs of social networks.

The vulnerability of heterogeneous networks to epidemics is bad news, but understanding it can provide good ideas for containing diseases. Ideally, almost all the population should be immunized to block the infection completely. However, if we can immunize just a fraction, it is not a good idea to choose people at random. Most of the times, choosing at random implies selecting individuals with a relatively low number of connections. Even if they block the disease from spreading in their surroundings, hubs will always be there to put it back into circulation. A much better strategy would be to target hubs. Inmunizing hubs is like deleting them from the network, and the studies on targeted attacks show that eliminating a small fraction of hubs fragments the network: thus, the disease will be confined to a few isolated components. This strategy faces a practical problem: nobody really knows the full map of social connections of a human group, so it is hard to identify its hubs. However, a clever tactic to find them was suggested in 2003 by physicists Reuven Cohen, Shlomo Havlin, and Daniel ben-Avraham: they suggested selecting people at random and asking them the name of somebody they are connected with. The most repeated names in this list are most likely the hubs of the social network: in fact, for its many connections, a hub will be tied to many people, so it will probably be mentioned by many of those interviewed. It should be noted that immunizing hubs works perfectly in theory, but many real-world details could impair it, such as whether the network disposes of specially redundant paths that go around hubs, or whether the network of contacts is fixed in time or evolving: for example, if Alice has HIV, and has unprotected sex with Bob, and Bob has unprotected sex with Carol, it makes a big difference to Carol whether Bob has sex with her before or after having sex with Alice.

The picture of the spread of an epidemic in a social network can be partially generalized to the case in which nodes do not represent people but locations (say, airports), and what spreads on the network are people (say, infected or healthy travellers). In this case, one can define a *global invasion threshold*, above which the disease becomes a pandemic, and below which it remains contained at local level. Closing airports is rarely a good idea: we would need to shut down 90 per cent of airports to block certain epidemics effectively, which would have too high a social and economic cost. Cleverer strategies, such as sharing antivirals with developing countries (which are often the source of new pandemics), are much more effective.

Viruses, ads, and fads

A couple of obscure Pakistani programmers, a university professor, a group of high school students... these were the authors of the first computer viruses. During the eighties, these parasite programs started to jump from one computer to another, basically hiding in the floppy disks interchanged by users. The first viruses were academic experiments on self-replicating software, but soon they escaped from the lab. In 1986, the *Brain* virus appeared from Pakistan. In the same year, a German laboratory lost control of *Virdem*. One year later, a group of students put *Vienna* in circulation. In the nineties, computer viruses were already a global problem, but this was nothing compared to what was in store.

The advent of the Internet brought a new generation of viruses that were capable of sending themselves to other computers through the Net. In 1999, *Melissa* spread through the Internet: people started to receive email messages with subjects such as 'Important message for you' or 'Here is the document you asked for... don't show anyone else ;-)'. The mail contained a file called list.doc. If the receiver opened it, it launched a program that sent the same message to the first fifty email addresses held in the computer. *Iloveyou*, *Slammer*, *Sobig*, *Blaster*, and many others exploded across the Net, using similar mechanisms and with disastrous effects: some of them destroyed companies' computer systems, universities' databases, and even affected Internet traffic.

Some features of a computer virus infection are strikingly similar to real-world epidemics. A computer becomes infected through its connections (for example, the social contacts of its owner as sampled by his or her email network) and infects others similarly. Some of the conclusions reached for diseases explain the puzzling behaviour of computer viruses. Even if antivirus programs are quickly updated, some viruses still circulate years after their first strike. This is no surprise if one considers the features of an epidemic spreading in scale-free networks: even if a large proportion of computers are immunized through antivirus programs, this is not enough to eradicate the infection: there is always some high-degree node here or there that puts it back in circulation.

This characteristic of endurance, which is a real problem with computer viruses, can be turned into a resource if one wants to disseminate information in a heterogeneous network. This is the principle behind *viral marketing*. Thanks to virtual social networks, today the WWW is full of videos, games, and applications that have 'gone viral': they are being forwarded by hundreds of thousands of people to all their contacts every day. One of the first examples of this idea was the spread of the Hotmail email service. In 1996, the company inserted into emails an automatic footer saying 'Get your free web-based email at

Hotmail', containing a link to a form for setting up a new email address in a few seconds (see page 50). Similar strategies were implemented by the email services of Yahoo! and Google, and by many social networks that are launched on the basis of providing access by invitation only.

Viral marketing takes advantage of an underlying psychological phenomenon called *social spreading*. This is the general tendency of people to mimic their contacts' behaviours, and to spread gossip, fads, rumours and ideas. This mechanism also acts in innovation adoption, group problem solving, and collective decision making. Sociologists and psychologists have found many examples of the striking tendency of humans to 'copy' each other. In 1962, a group of girls at a mission school in Tanzania experienced an unusual tendency to uncontrollable laughter. After a few months, tens of pupils of the school showed the same symptoms, and other people in the villages where some of the pupils were sent to rest showed the same disquieting giggling. After much investigation, doctors A. M. Rankin and P. J. Philip, who studied the case, came to the conclusion that it was an instance of 'mass hysteria'. A similar case was reported in 1998 in a high school in Tennessee, when the experience of a teacher who had the feeling of smelling gasoline spread to hundreds of students. All environmental factors were excluded, and scientists came to the conclusion that a kind of 'emotional contagion' was at work.

Many similar cases of social spreading have been documented, but in recent years scientists have found that the same mechanism may play a role in less exceptional settings: for example, obesity and smoking seem to spread on social networks. Three reasons are behind the fact that people connected share certain features or behaviours. First, there are external factors, such as belonging to the same social class: for example, people belonging to lower social classes have an increased risk of smoking and becoming obese; at the same time they are more likely to establish ties with each other than with people of a higher social class. Second, there is homophily: people that smoke or have similar body mass tend to make friends with those with similar habits. Third, there is social spreading: if you are a friend of somebody who smokes or is overweight, you are more likely to consider taking up smoking or increasing your daily food intake. Probably all three mechanisms are at work, but social spreading is likely to be the least trivial of them and should not be underestimated. Sociologists argue that it's not a specific condition that spreads; rather, it's the sharing of norms about what is appropriate that is disseminated. This perspective could be used in public health, to foster safer habits by targeting hubs in social networks.

Naturally, the contagion of behaviours, as well as rumours and ideas, is different in many respects from that of diseases. Unlike contagion, the act of spreading information is necessarily intentional. On the other hand, acquiring information is usually advantageous, so it is a more active process than getting infected. Learning or being convinced may need a longer exposure than getting a disease. Moreover, many other competing mechanisms are present. If social spreading was the leading factor, uniformity would be the rule, but in fact mechanisms against simple assimilation generate diversity, minorities, and polarization. In any case, in certain settings social contagion may indeed be the most relevant mechanism. In the forties, Richard Feynman invented *Feynmann's diagrams* as tools for modern high-energy physics. Some physicists accepted them with enthusiasm and others with scepticism, but they finally triumphed. A study of their diffusion in the communities of physicists of the US, Japan and USSR revealed that the observed trends could be quite accurately fitted with models used for epidemics, provided that the parameters were tuned to very different values.

Which came first, networks or dynamics?

One of the keys to the success of ancient Rome was its strategic position close to the River Tiber, which at the time was first and foremost a communication and commercial route. When the city became more powerful, it started building the first branches of its formidable road network. In their turn, the roads were a crucial tool for maintaining and further expanding Rome's power, since they provided a quick way to move goods and legions. More roads meant more power, and more power made it necessary to create even more roads: the result was that 'All roads lead to Rome,' according to an Italian saying. Similar patterns of development can be observed in almost every important city. A growing city attracts traffic and requires more connections (roads, railways, airlines...), which in their turn increase traffic and growth, which imply even more connections, etc. The communication network influences the dynamics of traffic, but this in turn reshapes the network, in a feedback loop.

Asking how network topology affects dynamics implies an assumption: that the network is an immutable structure, on top of which processes take place. In reality, all networks change *during* the dynamics. Therefore this assumption makes sense only if the timescale of the dynamics is much faster than that of the topology. This is reasonable for certain processes: for example, information interchanged on a daily or weekly basis spreads on a fixed social network, since usually friendship and kinship have turnover times in the order of years; or vehicle traffic towards a city during any given day moves on a fixed set of pathways: usually the street connections do not change every day.

However, in other cases this assumption is flawed. For example, in the epidemic spread of sexually transmitted diseases the timing of the links is crucial. Establishing an unprotected link with a person before they establish an unprotected link with another person who is infected is not the same as doing so afterwards. If we want to study the development of a city throughout a decade, it is necessary to take into account the interplay between traffic and changing connections. In certain technological networks, such as peer-to-peer file-sharing systems, network structure and information dynamics change on the same timescale and are strongly interwoven. In foodwebs, population dynamics can produce a reorganization of the network. When overfishing pushes a species below a certain level, the foodweb rearranges the predations and new species take the place of the old one. The coupling of network structure and dynamics is especially relevant at a time of virtual social networking. These tools provide a constant information flow about the structure and content of a person's social network. So researchers argue that this enhanced awareness may alter the way in which people create, maintain, and leverage their social networks.

Several approaches are possible to cope with the problem of coupling network structure and dynamics. For example, one can build network models through optimization, in which the quantity to be optimized is related to dynamics such as flow or searching. A more refined approach consists of modifying the fitness model in such a way that the value of the fitness depends on some dynamic parameter. When dynamics proceed, fitness changes accordingly; and this allows a reshaping of the networks. Other strategies are possible, all of them underpinning a basic idea: when a dynamic takes place in or is coupled to a network structure, then most of the time it is essential to take into account the underlying graph to fully understand what is going on.

Chapter 9
All the world's a net; or not?

One of the fathers of quantum mechanics, Paul Dirac, was reported to say about the revolutionary discoveries of physicists at the beginning of the 20th century: 'the rest is chemistry'. He meant that all science could be derived from the first principles of physics. Unfortunately, no more than a few cases, essentially the atoms of hydrogen and helium, can be solved precisely with quantum mechanical equations. More complex objects, like molecules, must be approached through approximations or computer simulations. Apart from a few macroscopic quantum effects, at the moment fundamental physics is relatively useless in understanding biology, the mind, or society. Similar mistakes occur in genetics, when DNA is incorrectly framed as something that can explain all the features, diseases, and behaviours of humans. In general, the results of basic science should not be taken beyond their real range of effectiveness, and it should be acknowledged that more specialized disciplines can give much deeper insights beyond that range. Network science should avoid the trap of overhyping. Its holistic vision, the revelation of unexpected similarities between widely different systems, and the current cultural fascination with the concept of network bring with them the temptation to think of network science as a 'theory of everything'. Sociologists, engineers, biologists, and philosophers have warned about the bald generalizations drawn from network theory. Most of this criticism is reasonable, but the results of network science should not be underestimated, nor its potential for future discoveries impaired.

The first major limitation of network science is the hunger for large-scale data. Methods used in social science, such as questionnaires and interviews, are costly, time consuming, and sometimes prone to subjective biases. Data drawn from information technologies (phone calls, email, social networks, geo-localization, RFID chips, health data, credit cards, etc.) provide unprecedented access to people's social relations, but they pose some problems too. A person delivering pizzas receives many phone calls, but most of them are from clients, not from friends: the *pizza delivery guy problem* shows that sorting out relevant information from large amounts of IT data (the list of phone calls, in this example) is not easy. Moreover, it should be mentioned that data mining with networks also creates ethical problems related to privacy and to their use by the military.

In many situations, only partial data are available. In order to draw aquatic foodwebs, ecologists capture fish and examine their digestive tracts: with this method, even the most brilliant scientist can miss some of the links. Genetic methods to deduce physical interactions between proteins can produce both false positives and false negatives. Maps of the Internet and the WWW are obtained by sending a 'probe' from a node to explore the surrounding edges: a sufficient number of paths can give a fairly good representation of the network, but some edges may never be discovered by the probes.

Once the data are available, representing them in a graph is inevitably a reductionist act. The focus on topology is one of the strengths of the network approach, but it forgets many of the specific features of the elements. If we are interested in those features, the graph approximation can be inadequate. Sometimes, but not always, network models can be modified to include these features.

The graph representation can fall short when geography (that is, the physical location of nodes) is more important than topology. For example, the position of electrical substations, airports, or train stations is obviously relevant to the arrangement of their connections. Furthermore, in social networks and foodwebs, closeness can determine the actual possibility of establishing a relation. Another element that may escape graph representations is time. For example, in sexually transmitted diseases, establishing a link with a person *after* he or she is infected by another person is obviously very different from doing so *before* the infection: the timing of links is crucial for the spread of the disease. Timing is important in a large range of networks: for example, scientific publications are arrayed in time, and they can only cite papers that came out in the past.

Sometimes, identifying nodes and edges is not trivial. It is easy to distinguish between an eagle and a hawk, but one could easily lose count of the number of bacteria in an ecosystem. To avoid underestimating the number of 'small' species, ecologists customarily aggregate organisms in *trophic species*, sets of organisms that share the same predators and the same preys. Similarly, social scientists aggregate individuals that are *structurally equivalent*: people that have the same number and kind of ties, for example within a family group. Similar procedures are applied to the Internet at the autonomous system level, the network of brain areas, etc. These procedures have to be performed in a coherent way to obtain a network that makes sense. Defining edges can be even more complicated. A company can hold a small fraction of the capital of another one, or up to 100 per cent of it. Two airports can be connected by one flight per day or by one per hour. In all these cases, a threshold must be established, below which a relation is considered too weak to be recorded. Weighting or putting a threshold on links strongly influences the shape of the resulting network, and must be done with very good reasons.

Once data have been arranged as a graph, a careful interpretation of the results is essential. An entire branch of network science is devoted to visualization, that is, producing algorithms to arrange nodes and links sensibly on paper or computer screen. However, most of the conclusions cannot be drawn by visual inspection, and mathematical analysis is needed. Criticisms have been made on the basis that not all complex networks have a heterogeneous degree distribution (and in any case this is never a mathematically exact power law). It is true that networks can be interesting also if they are not heterogeneous, but it is fair to say that often interesting networks are indeed heterogeneous. Perfect power laws are not vital: what matters is the presence of fat tails, revealing the existence of hubs. Interpreting heterogeneity as a signal of self-organization has been criticized, pointing out that a certain level of planning is present in many networks, as in the case of the precise design of local networks by administrators in the Internet. However, there is no doubt that the Internet, like many other networks, has not been designed at the large-scale level. So it is reasonable to argue that their deviations from randomness may be attributed to self-organization. Moreover, degree heterogeneity is just one signal of complexity in networks. Heterogeneity is also found in other features, such as their betweenness, clustering, weights, etc., and complexity is visible as well in other features, different from heterogeneity: for example, the modularity and community structure, which often deviates from randomness and has strong effects on the dynamics that take place on top of networks.

Another criticism points to the fact that network science has found nothing more than fuzzy similarities between different systems and dynamics, but not real *universality classes*. These are groups of different phenomena that correspond to the same basic mathematical laws (once certain details are discounted). Certainly, the specific features of a biological network are totally different from those of a technological one, and the diffusion of a computer worm has different rules from the contagion of a disease. However, network theory provides a framework of shared trends and common predictions for such different structures and processes. Usually, if the system is big enough and the phenomena are observed for long enough, they will display fairly similar trends.

Network science has shown predictive power in several fields. It is currently being used in consulting, to help organizations to better exploit the skills distributed across its members; in public health, to predict and prevent the spread of infectious diseases; in police and the military, to track terrorist, criminal, or rebel networks; and in several other fields. There are many problems to be tackled: among them, making more detailed models to fit specific networks and dynamics; finding new relevant network data; digging deeper in topology to find unnoticed regularities and fully explain the existing ones; characterizing small networks and learning how to deal with networks of networks; connecting biological networks more effectively with the evolutionary paradigm; finding new applications (for example in drug design); and possibly finding universality classes.

A famous quote from Galileo Galilei states: 'Philosophy is written in this grand book, the universe... it is written in the language of mathematics, and its characters are triangles, circles, and other geometric figures....' We believe that, especially in our complex contemporary world, we now need the 'characters' of networks.

Further reading

Popular science books on networks

Albert-László Barabási, *Linked, The New Science of Networks*. Perseus, New York (2002)

Mark Buchanan, *Nexus: Small Worlds and the New Science of Networks*. W. W. Norton & Co, New York (2002)

Ricard Solé, *Redes complejas. Del genoma a Internet*. Tusquets, Barcelona (2009)

Nicholas Christakis, James Fowler, *Connected. The Surprising Power of Our Social Networks and How They Shape Our Lives*. Little, Brown, New York (2009)

Introductory academic books on networks

Guido Caldarelli, *Scale-Free Networks*. Oxford University Press, Oxford (2007)

Romualdo Pastor-Satorras, Alessandro Vespignani, *Evolution and Structure of the Internet. A Statistical Physics Approach*. Cambridge University Press, Cambridge (2004)

Alain Barrat, Marc Barthelemy, Alessandro Vespignani, *Dynamical Processes on Complex Networks*. Cambridge University Press, Cambridge (2008)

L. C. Freeman, *The Development of Social Network Analysis: A Study in the Sociology of Science*. Empirical Press, Vancouver (2004)

Stanley Wasserman, Katherine Faust, *Social Networks Analysis. Methods and Applications*. Cambridge University Press, Cambridge (1994)

Albert-László Barabási et al. Interactive Network Science textbook project http://barabasilab.nev.edu/networksciencebook/ (in progress)

Articles and reviews

Mark Newman, Albert-László Barabási, Duncan J. Watts, *The Structure and Dynamics of Networks*. Princeton University Press, Princeton (2006) (includes bibliographic essays and an anthology of seminal papers in network theory)

Réka Albert, Albert-László Barabási, Statistical mechanics of complex networks. *Review of Modern Physics*, 74, 47–97 (2002)

M. E. J. Newman, The Structure and Function of Complex Networks. *SIAM Review*, 45 (2), 167–256 (2003)

S. Boccaletti et al. Complex Networks: Structure and Dynamics. *Physics Reports*, 424 (4–5), 175–308 (2006)

Katy Börner, Soma Sanyal, Alessandro Vespignani, Network Science. *Annual Review of Information Science and Technology*, 41 (1), 537–607 (2007)

Stephen P. Borgatti et al. Network Analysis in the Social Sciences. *Science*, 323 (5916), 892–95 (2009)

"牛津通识读本"已出书目

古典哲学的趣味	福柯	地球
人生的意义	缤纷的语言学	记忆
文学理论入门	达达和超现实主义	法律
大众经济学	佛学概论	中国文学
历史之源	维特根斯坦与哲学	托克维尔
设计,无处不在	科学哲学	休谟
生活中的心理学	印度哲学祛魅	分子
政治的历史与边界	克尔凯郭尔	法国大革命
哲学的思与惑	科学革命	民族主义
资本主义	广告	科幻作品
美国总统制	数学	罗素
海德格尔	叔本华	美国政党与选举
我们时代的伦理学	笛卡尔	美国最高法院
卡夫卡是谁	基督教神学	纪录片
考古学的过去与未来	犹太人与犹太教	大萧条与罗斯福新政
天文学简史	现代日本	领导力
社会学的意识	罗兰·巴特	无神论
康德	马基雅维里	罗马共和国
尼采	全球经济史	美国国会
亚里士多德的世界	进化	民主
西方艺术新论	性存在	英格兰文学
全球化面面观	量子理论	现代主义
简明逻辑学	牛顿新传	网络
法哲学:价值与事实	国际移民	自闭症
政治哲学与幸福根基	哈贝马斯	德里达
选择理论	医学伦理	浪漫主义
后殖民主义与世界格局	黑格尔	批判理论

德国文学	儿童心理学	电影
戏剧	时装	俄罗斯文学
腐败	现代拉丁美洲文学	古典文学
医事法	卢梭	大数据
癌症	隐私	洛克
植物	电影音乐	幸福
法语文学	抑郁症	免疫系统
微观经济学	传染病	银行学
湖泊	希腊化时代	景观设计学
拜占庭	知识	神圣罗马帝国
司法心理学	环境伦理学	大流行病
发展	美国革命	亚历山大大帝
农业	元素周期表	气候
特洛伊战争	人口学	第二次世界大战
巴比伦尼亚